CLASSICAL AND EVOLUTIONARY ALGORITHMS IN THE OPTIMIZATION OF OPTICAL SYSTEMS

Genetic Algorithms and Evolutionary Computation

Consulting Editor, David E. Goldberg
University of Illinois at Urbana-Champaign
deg@uiuc.edu

Additional titles in the series:

Efficient and Accurate Parallel Genetic Algorithms, Erick Cantú-Paz ISBN: 0-7923-7221-2

Estimation of Distribution Algorithms: *A New Tool for Evolutionary Computation,* edited by Pedro Larrañaga, Jose A. Lozano   ISBN: 0-7923-7466-5

Evolutionary Optimization in Dynamic Environments, Jürgen Branke   ISBN: 0-7923-7631-5

Anticipatory Learning Classifier Systems, Martin V. Butz   ISBN: 0-7923-7630-7

Evolutionary Algorithms for Solving Multi-Objective Problems, Carlos A. Coello Coello, David A. Van Veldhuizen, and Gary B. Lamont   ISBN: 0-306-46762-3

OmeGA: A Competent Genetic Algorithm for Solving Permutation and Scheduling Problems, Dimitri Knjazew ISBN: 0-7923-7460-6

The Design of Innovation: *Lessons from and for Competent Genetic Algorithms,* David E. Goldberg ISBN: 1-4020-7098-5

Noisy Optimization with Evolution Strategies, Dirk V. Arnold   ISBN: 1-4020-7105-1

Genetic Algorithms and Evolutionary Computation publishes research monographs, edited collections, and graduate-level texts in this rapidly growing field. Primary areas of coverage include the theory, implementation, and application of genetic algorithms (GAs), evolution strategies (ESs), evolutionary programming (EP), learning classifier systems (LCSs) and other variants of genetic and evolutionary computation (GEC). Proposals in related fields such as artificial life, adaptive behavior, artificial immune systems, agent-based systems, neural computing, fuzzy systems, and quantum computing will be considered for publication in this series as long as GEC techniques are part of or inspiration for the system being described. Manuscripts describing GEC applications in all areas of engineering, commerce, the sciences, and the humanities are encouraged. http://www.wkap.nl/prod/s/GENA

GENAGENAGENA
GENAGENAGENA
Genetic Algorithms and
Evolutionary Computation

CLASSICAL AND EVOLUTIONARY ALGORITHMS IN THE OPTIMIZATION OF OPTICAL SYSTEMS

by

Darko Vasiljević

Military Technical Institute, Yugoslavia
Military Academy, Yugoslavia
Mechanical Engineering Faculty, University of Belgrade, Yugoslavia

SPRINGER SCIENCE+BUSINESS MEDIA, LLC

Library of Congress Cataloging-in-Publication Data

Vasiljevic, Darko, 1960-
   Classical and evolutionary algorithms in the optimization of optical systems / by Darko Vasiljevic.
      p. cm. -- (Genetic Algorithms and Evolutionary Computation ; 9)
   Includes bibliographical references and index.
   ISBN 978-1-4020-7140-9   ISBN 978-1-4615-1051-2 (eBook)
   DOI 10.1007/978-1-4615-1051-2
      1. Optical instruments--Design and construction. 2. Lenses--Design and construction.
   3. Optics--Mathematics. 4. Genetic algorithms. I. Title. II. Series.

QC372.2.D4 V37 2002
681'4--dc21   2002069474

To Rada

Contents

Preface

The optimization of optical systems is a very old problem. As soon as lens designers discovered the possibility of designing optical systems, the desire to improve those systems by the means of optimization began. For a long time the optimization of optical systems was connected with well-known mathematical theories of optimization which gave good results, but required lens designers to have a strong knowledge about optimized optical systems. In recent years modern optimization methods have been developed that are not primarily based on the known mathematical theories of optimization, but rather on analogies with nature. While searching for successful optimization methods, scientists noticed that the method of organic evolution (well-known Darwinian theory of evolution) represented an optimal strategy of adaptation of living organisms to their changing environment. If the method of organic evolution was very successful in nature, the principles of the biological evolution could be applied to the problem of optimization of complex technical systems.

This monograph offers a unique presentation of research results in the field of classical and evolutionary algorithms and their application to the optimization of optical systems. Although, several good books about theory and practice of evolutionary algorithms and other optimization algorithms exist, (as well as books on the optical system design), this is the first book that covers optimization algorithms (both classical and modern evolutionary), demonstrating their application to the optimization of optical systems. This monograph shows the development of the author's research interests from the classical least squares method of optimization and its numerous modifications to the modern optimization methods that are based on analogies with nature-like genetic algorithms and evolution strategies. Every optimization method is completely described with the all-necessary mathematics in the first part of the book, and it is shown how to apply it to the optimization of optical systems in the second part. It is very clearly presented how an optimization method, which is a pure mathematical theory, can be implemented in the program for design and optimization of optical system.

This monograph is intended for:
- researchers and post graduate students in the field of optical design who wish to find out more about new and fascinating optimization methods such as genetic algorithms and evolution strategies;
- everyone who wishes to learn more about classical and evolutionary optimization methods in general and their application to the optimization of optical systems.

In this monograph one can find all necessary theories and practical implementation of these theories so it can be interesting for researchers from other fields, not only from optical design. In order to read this monograph one must have necessary mathematical and computer programming knowledge that can be obtained at any technical university.

The monograph is divided into four self-contained, logical units. In the introduction chapter some necessary terms are defined.

Part I is dedicated to the mathematical foundations of classical and evolutionary algorithms in the optimization of optical systems. All necessary mathematical theories are presented here. In Chapter 2 one can find a complete description of the least squares method and various kinds of damped least squares that tried to improve some parts of the basic optimization algorithm. There are also two other optimization methods (Spencer and Glatzel) which are interesting, but did not get much popularity in the optimization of optical systems. Chapter 3 deals with genetic algorithms. At the beginning of the chapter, an introduction and general description of the genetic algorithms are given. There are two main types of genetic algorithms: the binary genetic algorithm and the continuous parameter genetic algorithm. Every genetic algorithm has the following parts which are discussed: merit function scaling, selection methods and genetic operators. At the end of the chapter, some especially interesting genetic algorithms are described. In Chapter 4 evolution strategies are presented. At the beginning of the chapter an introduction and general description of the evolution strategies are given. In the remaining part of the chapter, the detailed description of two membered evolution strategies (variant ES EVOL) and multimembered evolution strategies (variants ES GRUP, ES REKO and ES KORR) are given. Chapter 5 deals with the comparison of the presented optimization algorithms and research possibility of hybridization of optimization algorithms by joining good features of classical and genetic algorithms.

Part II of this monograph is dedicated to the optical design fundamentals. Only things necessary for the application of optimization algorithms in the optimization of optical systems are presented here. Chapter 6 deals with different ways of the calculation of ray trace through the optical system. Raytracing is very important for calculation of different properties of optical systems. In this monograph paraxial, real and astigmatic ray traces are discussed. In Chapter 7 the aberrational theory is presented. There is no ideal optical system and every optical system has smaller or greater errors i.e. difference from an ideal optical system. These errors are called aberrations and the goal of every optimization algorithm is to try to minimize these aberrations. Part I and Part II represent the theoretical part of this monograph with a lot of mathematics, which is necessary to define a problem completely and to show possible ways of solution. Part III and IV represent the practical part of the monograph showing how this mathematical theory can be implemented in a real problem of the optimization of optical systems as well as results of various optimizations.

Part III deals with the implementation of the classical and evolutionary algorithms in the program for the optimization of optical systems. Every chapter has a detailed description of the optimization method with the flow chart diagram so a reader can see how the mathematical theory is turned into procedures. Chapter 8 deals with the implementation of the classical damped least squares (DLS) method. Chapter 9 presents the implementation of the adaptive steady-state genetic algorithm which was chosen among several genetic algorithms. Chapter 10 deals with the implementation of the two membered evolution strategy variant ES EVOL. The multimembered evolution strategies variants ES GRUP and ES REKO are presented in Chapter 11. And finally, the multimembered evolution strategies variant ES KORR is described in Chapter 12.

Part IV presents the results and the discussion of optimization of various types of objectives. In this monograph results of the great number of optimizations are presented for the following common objectives: the Cooke triplet, the Petzval and the Double Gauss.

Darko Vasiljević
Belgrade, Yugoslavia

Acknowledgements

I am very grateful to my colleagues Kenneth E. Moore from the Focus Software inc. and Professor Greg W. Forbes from the Department of Physics, Macquarie University, Sidney, Australia, for reviewing my manuscript and offering comments on this work.

Special thanks are due to Jasna Višnjić English language translator in the Military Technical Institute and a visiting professor for English at the Military Academy for her precious assistance with the English writing style.

I would like to thank Professor David E. Goldberg from the University of Illinois, a consulting editor for the Kluwer International Series on Genetic Algorithms and Evolutionary Computation, for reviewing my manuscript.

I would like to extend my thanks to the editors of the Kluwer Academic Publishers: Lance Wobus and Melissa Fearon for their help and patience on publishing and editorial matters.

Chapter 1

Introduction

1.1 Historical overview of optical system design

The optical design is one of the oldest branches of phisics, with a history going back over three centuries ago. In order to be able to design optical systems it is neccessary to be able to calculate the path of rays through an optical system. The basic requirements for the ray tracing is the knowledge of the law of refraction and the Cartesian coordinate geometry and the tables of logarithms if more advanced computing aids are not available. Suprisingly, most of these have become available within a short time. Napier published the tables of logarithms in 1614, Snell described the law of refraction in 1621 and Descartes published his famous Geometry in 1637. The first ray tracing is done probably by Gascoigne who died in 1641 at the age of 24. His work is not published but it is known to Flamsteed, the first English Astronomer Royal.

In order to use ray tracing in optical design it is necessary to be able to accurately measure the refracting surfaces radii of curvatures and the refractive indices of optical materials and this was not possible until the nineteenth century. The ray trace was not used because of large number of necessary calculations and relatively inaccurate input data. The optical design was carried out experimentally by making an optical system and then testing it. After necessary tests the experience is used for the optical system alternation. This situation continued until the introduction of the primary aberration theory in the forms which were derived from the work of Seidel. Early forms of this theory were rather cumbersome and not of much use to optical designers. The primary aberration theory was gradually simplified until, by the end of the nineteenth century, it was of great value and was used by many optical designers in their regular work. One of them was Taylor who used the primary aberration theory to design a famous Cooke triplet in 1893.

Even at this stage in the evolution of optical design methods, ray tracing was not used to obtain a comprehensive picture of the properties of new optical systems. For a number of reasons it was quicker and more economical to make an optical system and to test it, than to calculate its aberrations by tracing a large number of rays through the optical system.

In the early part of the twentieth century, ray tracing gradually became easier, as mechanical computing machines became generally available. The ray tracing is gradually becoming more and more important. During the second world war the most of optical manufacturers hired the large number of employees, mostly girls, to carry out ray tracing. A typical time for tracing a ray through a single optical surface by using mechanical calculating machines for a skilled worker is 10 minutes. If an average optical system has twelve surfaces then two hours were needed to trace just one ray through the whole optical system. To trace twenty rays for this optical system the whole working week of an employee is needed. This was barely acceptable and there was an obvious temptation for an optical designer to try to manage with an insufficient number of rays. It appears that twenty rays are roughly the number of rays needed to evaluate an optical system with a moderate aperture and moderate field angle.

The announcement of the Eniac, the first electronic computer in 1946 made a spectacular enterance into scientific calculation in general and the ray tracing in particular. During the period 1946 to 1960 there was a rapid progress in electronic computers development. Some of early computers made mistakes and the optical designers were often asked to check out computers because they knew what was needed in ray trace and how to do it. They were so familiar with the results from the ray tracing that they were quick to detect machine errors. The optical design programs were used on a nearly every new computer developed. The IBM 650 was the first really reliable computer. It changed programmers' debugging methods from suspecting a machine error to suspecting a software error.

The general introduction of computers into the optical industry was rather slow because of the way they were used. The computers were very fast (typical time for tracing a ray through a single optical surface is one second) but they were standing idle for a large proportion of the time waiting for the user to enter the input data. The obvious solution to the inefficient use of the computers was to devise optimization procedures which would allow the computer to make the required changes to the design parameters of the optical system. It is well known, from all of the experience obtained before the introduction of computers, that it is never possible to make all the aberrations (image errors) zero. The best that can be done is to minimize the aberrations until they are small, but finite. If the aberrations are small enough, the optical system will give an acceptible performance, and if they are not suffciently small, the optical system obviously becomes unacceptable. All serious attempts to use computers in optical system optimization have recognized this, but they differ in methods they employ to overcome this problem. This monography describes various classical and modern optimization methods.

From 1960 to 1980 the optical design and optimization programs were written in FORTRAN and were executed on mainframe computers. These optical design programs were large, expensive and user-unfriendly programs running on large expensive and slow by today's standards computers. With the invention of workstations with the Unix operating system and the X Windows graphical user interface the situation has dramatically changed. The graphical workstations were substantialy cheaper and more powerful than the mainframe computers because all resources of computer were used by only one designer and they were not shared among a great number of different users. The properties of graphical workstations quickly led to the idea that one workstation could be dedicated only to the optical design. From there to the idea of interactive optical design there was not a big step. The idea of interactive optical design set greater requirements before computer and programmers. With the introduction of PC computers with the Windows graphical operating system the history is once more repeated. A PC was much cheaper than a workstation. The graphical workstations are still applied only in small set of designers with the largest demands who do not think about the price. They only think about performance which must be of the first class.

Until the introduction of PC computers the optical design was almost exclusively the domain of dedicated and generally gifted professionals working in the classical optical design. With the introduction of PC computers and the powerful commercial programms for optical design, the optical design comunity was expanded with the professionals from other branches of technology. These people were not dedicated optical designers but used optical design to solve their problems.

1.2   Classical optical design stages

The classical optical design can be outlined in several stages:

Problem definition

Other engineers or disciplines may define the requirements and boundaries of the design problem, but it is up to the optical designer to translate these into detailed optical specifications and to define performance specifications. There may be many requirements to meet, some of them contradictory in terms of how to approach the design (e.g., large aperture and light weight are usually opposing requirements that lead to tradeoffs and compromises).

Pre-design

Once the basic properties of optical system are set, it is necessary to make a large number of decisions which will describe the way the optical system will be functioning. The decisions about the optical system that are made in this stage are of the following type:

− the optical system consists primary from lenses or from mirrors:
− the number of elements (lenses, mirrors and/or prisms) from which the optical system will consist:
− the overall size of the optical system.

The pre-design often involves paper and pencil sketching, including rough graphical ray tracing, with thin lenses standing in for real lenses. Usually in this stage it is necessary to check a large number of potential solutions to find on optimum solution.

Starting point selection

In this stage the concept moves towards reality, often with a help of an existing solution for a similar situation. Books, patents and optical designer's previous work are rich sources of information. The access to a database of existing designs can really speed up the selection process. Graphical and approximate methods can be also used to create a starting point "from a scratch" if necessary.

Initial analysis

It helps to have a baseline analysis of the optical system representing the starting point so that the improvement versus the specifications can be evaluated. The aberration analysis may not be a part of the specifications, but it will probably be useful in the design process, especially in selecting variables for optimization.

Optimization

At this stage the role of computers and optical design and optimization programs becomes irreplaceable. The first step in the optimization is to define:

– a set of variables (parameters such as curvature, thickness, index of refraction, etc. that the program can change in order to try to improve the performance of the optical system);

– a merit function (measure of optical quality, zero typically implying "perfection");

– constraints (boundary values that restrict possible configurations of the optical system).

The optimization is done with well known mathematical (numerical) methods which are used to alter the variables in systematic ways that attempt to minimize the merit function while honoring all constraints. Since the optimization is a very complicated process it is often necessary that optical designer guide the optimization process in a particular direction.

Final analysis

After the optimization method finds the optical system with the minimum merit function it is necessary to determine whether the found optical system satisfies all original specifications. This is very important because some specifications like the minimum resolution or the MTF cannot be included in the merit function. If any of the specifications are not fulfilled it is possible that an additional optimization is needed with a changed set of variable parameters. In some cases it may be necessary to start the optimization with a completely new optical system.

Preparation for Fabrication

When the optical system is designed to satisfy all specifications it is necessary to do a lot of things before the optical system can be fabricated. After the optical design for each optical system the mechanical design is needed. The mechanical design enables the optical elements to accomplish their function. It is very important to perform a detailed analysis of manufacturing error effects, which is often called a tolerance analysis.

1.3   The concept of system optimization

A human desire for perfection finds its expression in the optimization theory, which teaches how to describe and fulfil an optimum. The optimization tries to improve system performances in the direction of the optimal point or points.

The optimization can be defined as a part of the applied or numerical mathematics or a method for system design by computer in accordance with either one stress theoretical aspect (existence of the optimum solution conditions) or the practical aspect (procedures for obtaining the optimum solution).

The most general mathematical formulation of the optimization problem is:

Minimize $\qquad\qquad\qquad\qquad f(x)$

and fulfil boundary conditions $\qquad g_i(x) = b_i\,; \quad i = 1(1)\,m$ $\qquad\qquad\qquad$ (1.1)

$$h_j(x) \le c_j\,; \quad j = 1(1)\,n$$

where:

X is the vector of system construction variables;

$f(x)$ is the merit function;

$g_i(x),\ h_j(x)$ are general functions which define the boundary conditions;

m is the total number of equality boundary conditions;

n is the total number of nonequality boundary conditions.

A very important part of every optimization is the definition of the merit function. The idea of the merit function is basic to any design type. Suppose one has two designs A and B for a certain application and ought to decide which one is better and will be manufactured. One can write the following statements:

$\qquad\qquad$ Design A is preferred to design B $\qquad\qquad\qquad (A \supset B)$,

$\qquad\qquad$ Neither design A nor design B is preferred $\qquad\quad (A = B)$,

$\qquad\qquad$ Design A is less preferred than design B $\qquad\quad (B \subset A)$.

These expressions define the operation of preference. For these expressions the following laws are fulfilled:

reflexivity:

$$A \subset B \Rightarrow B \supset A \quad, \quad A = B \Rightarrow B = A \qquad\qquad (1.2)$$

transitivity:

$$\begin{matrix} A \subset B \\ A = B \end{matrix} \quad \text{and} \quad \begin{matrix} B \subset C \Rightarrow A \subset C \\ B = C \Rightarrow A = C \end{matrix} \qquad\qquad (1.3)$$

Any number of designs can be ordered by preference $A_1 \subset A_2 \subset A_3 \subset ... \subset A_n$. By this procedure each design in the collection is related to a positive integer. Such a relation can be considered as the merit function. In the other words the merit function is any function, which maps a collection of designs onto the positive integers according to a stated preference rule. The above definition can be generalized for the infinite case. The collection becomes infinite when one or more design parameters varies continuously over some interval. Then it is logical to change the integers with the real numbers. The definition of the merit function may be written formally as:

The merit function is any function, which maps the design space upon the set of positive real numbers in a manner consistent with the preference rule.

The merit function is simply a convenient way of expressing the preferences of the designer. The general definition of the merit function is given above but there is a problem of selecting a suitable form of the merit function. From the mathematical standpoint, the merit function is desirable to be:

– continuous;
– differentiable in all design parameters;
– easy to compute.

The most common choice for the merit function is the one used by Legendre in the classical method of least squares. If $(f_1, f_2, ..., f_m)$ are the set of system output values and $(\omega_1, \omega_2, ..., \omega_m)$ are the set of positive weighting factors then the merit function is defined by the equation:

$$\psi = (\omega_1 \cdot f_1)^2 + (\omega_2 \cdot f_2)^2 + ... + (\omega_m \cdot f_m)^2 \qquad (1.4)$$

This type of merit function has many advantages. The quantity ψ is a positive number, which is zero only if a system is perfect. In every iteration there exist two designs - one is discarded and the other is accepted to become the initial design for the next iteration. In the ordinary course of design, such decisions must be made many times. The explicit use of the merit function supplies the program with the decision rule, which is in accordance with the previously expressed wishes of the designer.

1.4   The definition of the optical system optimization

Optical systems are defined by the parameters associated with the individual surfaces of the system. These parameters can be divided into two main groups:

– the basic parameters which must be defined for every optical system;
– the optional parameters which are defined only if this parameter exists, for example if the surface is aspheric the aspherical constant must be defined.

The basic parameters are the following parameters:

– the radius of curvature for each optical system surface;
– the separation between the optical system surfaces;
– the glasses which optical system components are made of;
– the clear radius for each optical system surface.

To define and analyse the optical system completely, one must, besides the basic parameters, define the ray data, which describe rays that are to be traced through the optical system.

The optimization concept is fairly simple but its implementation can be quite complex. It can be defined as given an initial system definition and the set of target performance goals determines the set of system configuration variables, which minimizes the deviation of actual performance from targeted goals without violating boundary limitations. This simply stated problem becomes a complex problem when the number of variables and targets is large, errors are non-linear functions of the variables, and the errors are not orthogonal to each other with respect to the variables. In a typical optical system optimization case, all of these conditions are true to varying degrees.

In a general case there are more aberrations than system configuration variables and most that can be done is minimization of the finite differences (residuals) of aberrations.

Let an optical system have N constructional parameters. Any ray passing through the system may be specified by six numbers, three specifying its coordinates and three specifying its direction cosines. From these six numbers only four numbers are free because the ray must fulfil the following conditions:

- the coordinates must satisfy the surface equation,
- the direction cosines must satisfy the relation:

$$L^2 + K^2 + M^2 = 1 \tag{1.5}$$

where:

K is the direction cosines of the ray along the x axis;

L is the direction cosines of the ray along the y axis;

M is the direction cosines of the ray along the z axis.

If coordinates (X_0, Y_0, L_0, K_0) are taken to define an object ray, the image ray, defined by four quantities (X_F, Y_F, L_F, K_F) at the image plane, can be computed as:

$$
\begin{aligned}
X_F &= X_F(X_0, Y_0, L_0, K_0, x_1, x_2, x_3, \dots, x_n) \\
Y_F &= Y_F(X_0, Y_0, L_0, K_0, x_1, x_2, x_3, \dots, x_n) \\
L_F &= L_F(X_0, Y_0, L_0, K_0, x_1, x_2, x_3, \dots, x_n) \\
K_F &= K_F(X_0, Y_0, L_0, K_0, x_1, x_2, x_3, \dots, x_n)
\end{aligned} \tag{1.6}
$$

where:

X_0, Y_0 are the ray coordinates along the x and y axis in the object space;

L_0, K_0 are the direction cosines along the x and y axis in the object space;

$x_1, x_2, x_3, \dots, x_n$ are the constructional parameters of the optical system.

The above equations represent a four-dimensional family of rays. If quantities X_0 and Y_0 are held fixed, a two parameter bundle of rays emerging from a particular object point is obtained. A requirement for sharp imagery is that this ray bundle should again coalesce to a point in an image space. In other words the ray coordinates X_F and Y_F ought to depend only upon the ray coordinates X_0 and Y_0 but not upon the ray direction cosines K_0 and L_0. Further, in order to secure freedom from distortion, so that geometrical similarity is preserved, the following relations must be specified:

$$X_F = M \cdot X_0$$
$$Y_F = M \cdot Y_0 \qquad\qquad (1.7)$$

where M is the optical system magnification. These equations can be writen by defining two aberration functions G and H :

$$G(X_0, Y_0, L_0, K_0, x_1, x_2, x_3, \dots, x_n) = X_F - M \cdot X_0$$
$$H(X_0, Y_0, L_0, K_0, x_1, x_2, x_3, \dots, x_n) = Y_F - M \cdot Y_0 \qquad (1.8)$$

These functions may be computed by ray tracing. The condition that a designed optical system is perfect is that the aberration functions G and H satisfy the relation $G \equiv H \equiv 0$ for all object ray values (X_0, Y_0, L_0, K_0) and some particular fixed values of constructional parameters $(x_1, x_2, x_3, \dots, x_n)$. This relation can be only nearly satisfied over a finite aperture in the (X_0, Y_0, L_0, K_0) space. So one ought to introduce the merit function to describe the quality of design. The merit function can take many different forms and the chosen form is of vital concern to the optical designer. After more than fifty years of research it is now clear that the merit function constructed by the sum of squares of aberration functions is optimal. It can be written:

$$\psi = \iiint (G^2 + H^2)\, dX_0\, dY_0\, dl_0\, dm_0 = \psi(x_1, x_2, \dots, x_n) \qquad (1.9)$$

where the integral is taken over a region in the (X_0, Y_0, L_0, K_0) space. Physically that means integration over a finite aperture and field. Since the aberration functions G and H are real functions, the merit function ψ is a positive number, which is zero if and only if both aberration functions G and H are identical and equal to zero. As the optical system quality decreases the merit function ψ increases, so the problem of the optical system optimization becomes a problem of the merit function minimization.

In practice the integration cannot be carried out analytically and the aberration functions G and H can be only evaluated by tracing a selected set of rays. Then the integral from Eq. (1.9) can be replaced by a finite sum:

$$\psi = \sum_{i=1}^{M} G_i^2 + H_i^2 \tag{1.10}$$

where each value of i represents a specific point $(X_0^i, Y_0^i, K_0^i, L_0^i)$. Since the distinction between the aberration functions G and H is not important, the problem description can be further simplified by writing:

$$\psi = \sum_{i=1}^{M} f_i^2(x_1, x_2, ..., x_n) \tag{1.11}$$

where f_i is one of the aberration functions G_i or H_i. The statement of the problem is still incomplete because the constructional parameters themselves are not entirely unconstrained. They ought to satisfy a set of inequalities of the form:

$$b_1(x_1, x_2, ..., x_n) \geq 0$$
$$b_2(x_1, x_2, ..., x_n) \geq 0$$

$$\cdot$$
$$\cdot \tag{1.12}$$
$$\cdot$$

$$b_n(x_1, x_2, ..., x_n) \geq 0$$

These boundary conditions ought to insure that:
— all optical elements can be manufactured;
— the distances between successive optical elements ought to be positive and greater than the predefined value;
— all optical elements ought to have sufficient centre and edge thicknesses;
— the glass variables are real glasses available in the glass catalogue.

The treatment of boundary conditions is an important part of an actual optimization program.

Chapter 2

Classical algorithms in the optimization of optical systems

The problem of the optical system optimization can be stated as follows:

Given an optical system with specified set of variables find a new optical system with set of variables, which improve performance of the optical system (merit function) subjected to physical and optical constraints.

All methods for the optical system optimization use a merit function for minimization. The merit function is usually formed from aberrations that can be expressed as functions of system construction parameters, the problem variables. In all cases of practical interest it is necessary to have more aberrations than variables to get good correlation between the optical system performance and the merit function being minimized by the optimization algorithm.

It is necessary to impose constraints on the problem definition. Constraints provide that all optical elements can be manufactured, the element centre and edge thickness are positive values greater than a predefined value, the glass variables are real glasses available in a glass catalogue.

Finally, a starting optical system ought to be provided by the designer as well as a list of constructional parameters (e.g. radii, thicknesses, glass indices of refraction and dispersions, etc.) which the optimization algorithm can use as variables. The starting optical system ought to be sufficiently close to the region where the solution to the problem is expected to lie. Furthermore the variables must be effective for reducing the merit function and satisfying the problem constraints.

The problem of the optical systems optimization is a non-linear problem because the merit function is the sum of squares of the aberrations which can be linear or more typically non-linear functions. In general, this problem cannot be solved and some approximations must be made. One of approximations is that the merit function will vary quadratically with respect to the variables if the aberrations are linear functions of variables. Over a limited region of variable movement this assumption is valid and leads to efficient algorithms based upon linearized solution algorithms.

2.1   Least squares optimization

Rosen and Eldert in [1] first proposed that the classical least squares method can be used in the optimization of optical systems. Since then great number of researchers used this method or some of its numerous modifications. The basis for appeal of least squares seems to be in the fact that the least squares method applied to the merit function preserves the information pertaining to the distribution of the various aberrations.

Let M be the number of aberrations, N be the number of the constructional parameters and $\mathbf{X}$ be a vector in multidimensional space of coordinates x_j. Let the aberrational function f_i be defined as a function of variable constructional parameters $f_i = f_i(x_1, x_2, ..., x_n)$. The function f_i is expanded in the Taylor series about a given point $(x_{01}, x_{02}, ..., x_{on})$ in the parameter space.

$$f_i = f_i(x_1, x_2, ..., x_n) = f_{0i}(x_1, x_2, ..., x_n) + \sum_{j=1}^{N} \frac{\partial f_i}{\partial x_j} \cdot (x_j - x_{0j}) + ... \qquad (2.1)$$

If the series is truncated at the linear term Eq. (2.1) becomes:

$$f_i = f_{0i} + \sum_{j=1}^{N} \frac{\partial f_i}{\partial x_j} \cdot (x_j - x_{0j}) \qquad (2.2)$$

where f_{0i} is the value of the function f_i for a given point $(x_{01}, x_{02}, ..., x_{on})$.

If the first partial derivative is designated by:

$$\frac{\partial f_i}{\partial x_j} = a_{ij} \qquad (2.3)$$

Then Eq. (2.14) becomes:

$$f_i = f_{0i} + \sum_{j=1}^{N} a_{ij} \cdot (x_j - x_{oj}) \qquad (2.4)$$

or in the matrix form:

$$\mathbf{F} = \mathbf{F_0} + \mathbf{A} \cdot (\mathbf{X} - \mathbf{X_0}) \qquad (2.5)$$

where:

F   is the vector of aberration functions;

F₀   is the vector of aberration functions about a given point $\mathbf{X_0}$ where functions are expanded to Taylor series;

X   is the vector of constructional parameters;

X₀   is the point in multidimensional vector space where aberrational functions are expanded in Taylor series;

A   is the Jacobian matrix – matrix of first partial derivatives of the ith aberrational function with respect to the jth constructional parameter.

Eq. (2.5) represents linear approximation for the aberrational functions vector **F** and it is valid only for low values $(\mathbf{X} - \mathbf{X_0})$. To simplify the prolem further on, the departure from initial zero value can be designated by **X** if the origin is shifted to $\mathbf{X_0}$. This shift of the origin reduces Eq. (2.5) to:

$$\mathbf{F} = \mathbf{F_0} + \mathbf{A} \cdot \mathbf{X} \tag{2.6}$$

The merit function is defined in the Chapter 1 by Eq. (1.11) as:

$$\psi = \sum_{i=1}^{N} f_i^2$$

$$\psi = \sum_{i=1}^{M} \left[f_{0i} + \sum_{j=1}^{N} \frac{\partial f_i}{\partial x_j} \cdot x_j \right]^2 \tag{2.7}$$

$$\psi = \sum_{i=1}^{M} f_{0i}^2 + 2 \cdot \sum_{i=1}^{M} \left[f_{0i} \sum_{j=1}^{N} a_{ij} \cdot x_j \right] + \sum_{i=1}^{M} \sum_{j=1}^{N} \sum_{k=1}^{N} a_{ij} \cdot a_{ik} \cdot x_j \cdot x_k$$

where x_j is the change of the jth constructional parameter.

In Eq. (2.7) the first term is constant and may be neglected. The second term may be made identical to zero by shifting the origin of x_j and changing the axes so that the merit function ψ can be expressed as a positive definite quadratic form ($a_{ij} \geq 0$ for each i, j). With this modification Eq. (2.7) can be written in the matrix form as:

$$\psi = \mathbf{X}^T \cdot \mathbf{A}^T \cdot \mathbf{A} \cdot \mathbf{X} \tag{2.8}$$

The level lines $\psi = \text{const}$, in N-dimensional hyperspace, are hyper-ellipsoids with lengths of their principal axes given by the eigen values of the matrix $\mathbf{A}^T\mathbf{A}$, which are determined in terms of first partial derivatives a_{ij}.

The properties of the matrix $\mathbf{A^T A}$ are very important for the least squares optimization and have been thoroughly mathematically researched. Jamieson in [2] gives detailed discussion about the properties of the matrix $\mathbf{A^T A}$ from which the following statements are taken.

For an arbitrary set of aberrations f_i the elements of Jacobian matrix $\mathbf{A}$ are not completely independent, but some aberrations can be expressed as a linear combination of other aberrations. The Jacobian matrix $\mathbf{A}$ is then singular and one or more its eigenvalues is zero. In practice the Jacobian matrix is not exactly singular but it is very nearly so. The matrix $\mathbf{A^T A}$ is even closer to singularity with a very large spread in the magnitudes of its eigenvalues.

The condition that the merit function ψ be minimum requires that the gradient is zero i.e. $\dfrac{\partial \psi}{\partial x_k} = 0$. Since $\psi = \sum\limits_{i=1}^{M} f_i^2$ solution is:

$$\frac{\partial \psi}{\partial x_k} = \sum_{i=1}^{M} 2 \cdot f_i \cdot \frac{\partial f_i}{\partial x_k} = \sum_{i=1}^{M} 2 \cdot \left[f_{0i} + \sum_{j=1}^{N} a_{ij} \cdot x_j \right] \cdot a_{ik} = 0$$

$$\sum_{i=1}^{M} f_{0i} \cdot a_{ik} + \sum_{i=1}^{M} \sum_{j=1}^{N} a_{ij} \cdot x_j \cdot a_{ik} = 0$$

or in matrix form:

$$\mathbf{A^T} \cdot \mathbf{F_0} + \mathbf{A^T} \cdot \mathbf{A} \cdot \mathbf{X} = 0$$
$$\mathbf{A^T} \cdot \mathbf{A} \cdot \mathbf{X} = -\mathbf{A^T} \cdot \mathbf{F_0} \tag{2.9}$$
$$\mathbf{X} = -\left(\mathbf{A^T} \cdot \mathbf{A}\right)^{-1} \cdot \mathbf{A^T} \cdot \mathbf{F_0}$$

Eq. (2.9) defines a classical least squares method, which was first applied in optics by Rosen and Eddert [1]. Jamieson in [2] points that the matrix $\mathbf{A^T A}$ is nearly singular matrix for all cases of practical interest. Elements of the solution vector $\mathbf{X}$ from Eq. (2.9) are then large. Any small error in calculation of the elements of the matrix $\mathbf{A^T A}$ i.e. calculation of the first partial derivatives or any round off error in the equation solving routine will seriously affect solution and this may be worse than the initial system. Another great problem is non-linearity of the aberrations in the vector $\mathbf{F}$. The solution of Eq. (2.9) is optimal only for strictly linear aberrations. However aberrations are in general non-linear functions and that means that the solution i.e. the optimum optical system can be worse then the starting optical system.

2.2   Damped least squares (DLS) optimization

In order to overcome difficulties in the classical least squares method it is necessary to limit changes of the constructional parameters so that there exist some reasonable correlation between the predicted improvement and the achieved improvement of the optical system. Kidger in [3] defines a value, which he calls a step length, for control and confinement of changes of the constructional parameters as:

$$\Phi = \sqrt{\sum_{j=1}^{N} x_j^2} \tag{2.10}$$

Wynne and Wormell in [4] suggest that the problem can be solved by following modification of the merit function:

$$\psi = \sum_{i=1}^{M} f_i^2 + p^2 \cdot \sum_{j=1}^{N} x_j^2 \tag{2.11}$$

where is p - scalar called a damping factor.

The solutions obtained from the least squares equations when the merit function is defined by Eq. (2.11) will be a compromise between the least squares solution and a solution which will keep the magnitude of the step length small. The exact compromise is determined by the value of the damping factor p.

The minimum of the merit function is:

$$\mathbf{A}^{\mathrm{T}} \cdot \mathbf{F_0} + (\mathbf{A}^{\mathrm{T}} \cdot \mathbf{A} + p^2 \cdot \mathbf{I}) \cdot \mathbf{X} = 0$$
$$\mathbf{X} = -(\mathbf{A}^{\mathrm{T}} \cdot \mathbf{A} + p^2 \cdot \mathbf{I})^{-1} \cdot \mathbf{A}^{\mathrm{T}} \cdot \mathbf{F_0} \tag{2.12}$$

where $\mathbf{I}$ is the unit matrix of N order.

The damping factor p is introduced to improve the reliability of the convergence process. The matrix $\mathbf{A}^{\mathrm{T}}\mathbf{A} + p^2\mathbf{I}$ is formed by adding p to every element in the main diagonal of the original matrix $\mathbf{A}^{\mathrm{T}}\mathbf{A}$. From the theory of linear algebra it is known that this increases every eigenvalue by p. This makes every eigenvalue positive and nonzero so that the matrix $\mathbf{A}^{\mathrm{T}}\mathbf{A} + p^2\mathbf{I}$ cannot be singular. The effect of this modification of least squares method is to limit the program in calculation of large changes of the constructional parameters (large iteration step length) when the improvement of the merit function is small. If the damping factor is great then small changes will be calculated in the constructional parameters, while if the damping factor is small then large changes will be calculated in the constructional parameters. If the damping factor is too small then changes of the constructional parameters will be out of linearity range and the optimization method will diverge the same as the least squares method.

The method of damping the least squares equations described above is known as "additive damping". The DLS method has the following very desirable properties:

− only first derivatives are needed;

− converges to a local minimum from a starting point far from solution (the global convergence);

− solution of Eq. (2.12) always exists if the damping factor p is nonzero;

− efficient linear algebraic processes can be used;

− can be implemented compactly.

Although these properties are very reassuring concerning the finding of a solution, the rate of convergence to the solution can be discouragingly slow. The efficiency of the convergence of the DLS method depends on the proper setting of the damping factor at the starting point of the optimization process. One of major problems with the DLS method is how to determine the damping factor when one has no special knowledge of it at the starting stage of optimization.

The rate of convergence of algorithm r is usually defined as:

$$r = \lim_{k \to \infty} \frac{\left\| \mathbf{X}^{k+1} - \overline{\mathbf{X}} \right\|}{\left\| \mathbf{X}^k - \overline{\mathbf{X}} \right\|^s} \tag{2.13}$$

where:

s is the positive integer order of convergence;

$\mathbf{X}^{k+1}, \mathbf{X}^k$ is the value of the constructional parameters vector $\mathbf{X}$ at the $(k+1)^{th}$ and the kth iteration;

$\overline{\mathbf{X}}$ is the value of the constructional parameters vector $\mathbf{X}$, which represents the solution.

The convergence rate is said to be quadratic for $s = 2$. If $s = 1$ the convergence rate is linear and if $s = 1$ and $r = 0$ then convergence rate is superlinear.

Meyer in [5] shows that the convergence rate for the DLS method is linear and related to relative sizes of the matrix $\mathbf{A}^T \mathbf{A}$ and the sum of products of the aberrations and the Hessian matrix ($\sum f_i \cdot H_i$). The Hessian matrix is defined as the matrix of second partial derivatives of the ith aberration f_i with respect to the vector of constructional parameters $\mathbf{X}$. The "size" of the matrix can be expressed in terms of its eigenvalues. Let m be the smallest eigenvalue of the matrix $\mathbf{A}^T \mathbf{A}$ and M the largest eigenvalue in the absolute value of $\sum f_i \cdot H_i$. If the ratio $\dfrac{M}{m} < 1$ then the linear rate of convergence is $r \leq \dfrac{M}{m}$. Mayer indicates three domains of performance:

– Zero residual case: if all the aberrations f_i are zero at the solution then the DLS converges at a superlinear rate to the solution.
– Small residual case: if the aberrations f_i and/or the Hessian matrix H_i are small compared to the matrix $\mathbf{A^T A}$ at the solution, then the ratio $\dfrac{M}{m}$ is much smaller then unity and the DLS converges at the fast linear rate to the solution.
– Large residual case: if the aberrations f_i and/or the Hessian matrix H_i are large compared to the matrix $\mathbf{A^T A}$ at the solution, then the ratio $\dfrac{M}{m}$ is of unity order and the arbitrary slow rate of linear convergence occurs.

2.3   Improvements of the damped least squares optimization

Numerous methods have been put forward to improve the convergence of the DLS in the large residual case. Simple alternative to the additive damping in the DLS is multiplicative damping as proposed by Meiron in [6]. The damping procedure in additive damping is unjustified in the sense that it treats all variables on equal basis, whereas in general the non-linearity or the sensitivity of the solution to derivative errors and round-off errors will vary from one variable to another. In additive damping the magnitude of the damping factor p will be determined by the least reliable variable. That is why Meiron introduces the following merit function:

$$\psi = \sum_{i=1}^{M} f_i^2 + p^2 \cdot \sum_{j=1}^{N} q_j^2 \cdot x_j^2 \qquad (2.14)$$

where the factors q_j are chosen according to the range of linearity of the constructional parameters x_j. Meiron defines the factors q_j as:

$$q_j = \sqrt{\sum_{i=1}^{M} a_{ij}^2} \qquad (2.15)$$

These factors are actually the absolute values of the M-dimensional vectors components of which are the elements of the jth column of the matrix $\mathbf{A^T A}$. By this definition a sensitive constructional parameter x_j, the one responsible for large first partial derivatives of the aberrations with respect to the constructional parameters is assigned a large damping factor, and consequently a small step is taken in that direction.

The minimum of the merit function is:

$$\mathbf{A}^{\mathbf{T}} \cdot \mathbf{F_0} + \mathbf{A}^{\mathbf{T}} \cdot \mathbf{A} \cdot \mathbf{X} + p^2 \cdot \mathbf{Q} \cdot \mathbf{X} = 0 \qquad (2.16)$$

The matrix $\mathbf{Q}$ is a diagonal matrix with elements q_j, which are equal to the principal diagonal terms of the matrix $\mathbf{A}^{\mathbf{T}}\mathbf{A}$. Then Eq. (2.16) becomes:

$$\mathbf{A}^{\mathbf{T}} \cdot \mathbf{F_0} + (1 + p^2) \cdot \mathbf{A}^{\mathbf{T}} \cdot \mathbf{A} \cdot \mathbf{X} = 0 \qquad (2.17)$$

The solution to this equation is:

$$\mathbf{X} = -\left[(1 + p^2) \cdot \mathbf{A}^{\mathbf{T}} \cdot \mathbf{A} \right]^{-1} \cdot \mathbf{A}^{\mathbf{T}} \cdot \mathbf{F_0} \qquad (2.18)$$

By comparing definitions of the merit functions given by equations (2.11) and (2.14) it can be seen that the additive damping is in fact special case for the multiplicative damping where the second part of the damping factor is $q_j = 1$ for all variables $j = 1(1)N$.

This procedure has, however, little theoretical justification. The damping factor q_j should be determined by the second partial derivatives, there is no a priori reason for assuming that a large first partial derivative implies a large second partial derivative. Notwithstanding the lack of sound theoretical basis both additive and multiplicative damping have been used with considerable success.

Buchele in [7] proposed an alternative way of the damping factor calculation for the DLS method, which has theoretical justification. Let the starting optical system has the constructional parameters defined in the vector $\mathbf{X_0}$ it is necessary to predict steps $\Delta\mathbf{X} = \mathbf{X} - \mathbf{X_0}$, which will result in a new optical system with the constructional parameters defined in the vector $\mathbf{X}$ for which the merit function ψ is minimum. If the merit function is minimum then the gradient is equal to zero and it is possible to use a well-known Newton – Raphson method.

$$\sum_{j=1}^{N} \left(\frac{\partial^2 \psi}{\partial x_k \partial x_j} \right) \cdot \Delta x_j = -\frac{\partial \psi}{\partial x_k} \qquad k = 1(1)N \qquad (2.19)$$

Substituting the merit function defined in the Chapter 1 by Eq. (1.11) in Eq. (2.19) the right side becomes in the matrix expression:

$$-\left[2\cdot\sum_{i=1}^{M}f_i\cdot\frac{\partial f_i}{\partial x_k}\right]=-2\cdot\mathbf{A}^\mathrm{T}\cdot\mathbf{F} \tag{2.20}$$

The left side of Eq. (2.19) becomes in the matrix expression:

$$\left[2\sum_{i=1}^{M}\left(\frac{\partial f_i}{\partial x_k}\frac{\partial f_i}{\partial x_j}+f_i\cdot\left(\frac{\partial^2 f_i}{\partial x_k\partial x_j}\right)\cdot\Delta x_j\right)\right]=2\mathbf{A}^\mathrm{T}\mathbf{A}\cdot\Delta\mathbf{X}+2\sum_{i=1}^{M}f_i\cdot\left(\frac{\partial^2 f_i}{\partial x_k\partial x_j}\right)\cdot\Delta x_j \tag{2.21}$$

The classical least squares method uses only first derivatives i.e. the first term in Eq. (2.21). Great inconvenience of the classical least squares is that the matrix $\mathbf{A}^\mathrm{T}\mathbf{A}$ is nearly singular and the vector $\Delta\mathbf{X}$ has large oscillating values. This behaviour can be corrected by adding the damping factor. Buchele shows in [7] that the second term in Eq. (2.21) provides damping but the number of the partial second derivatives is prohibitively great. So Buchele approximates Eq. (2.21) by neglecting the cross second partial derivates and calculating the homogeneous second partial derivatives. Then Eq. (2.21) becomes:

$$\left[2\cdot\sum_{i=1}^{M}f_i\cdot\left(\frac{\partial^2 f_i}{\partial x_j^2}\right)\cdot\Delta x_j\right]=2\cdot\mathbf{B}^\mathrm{T}\cdot\mathbf{F}\cdot\Delta\mathbf{X} \tag{2.22}$$

where the elements of the matrix $\mathbf{B}$ are $b_{ij}=\dfrac{\partial^2 f_i}{\partial x_j^2}$. These second partial derivatives can be computed as simply as first partial derivatives by using at most twice the computing time. These second partial derivatives are elements on the principal diagonal together with the elements f_i (the aberrations). Now the damped least squares equation becomes:

$$\left(\mathbf{A}^\mathrm{T}\cdot\mathbf{A}+\mathbf{B}^\mathrm{T}\cdot\mathbf{f}\cdot\mathbf{I}\right)\cdot\Delta\mathbf{X}=-\mathbf{A}^\mathrm{T}\cdot\mathbf{F} \tag{2.23}$$

Dilworth in [8] also uses second partial derivatives for damping least squares equations. Like Buchele he also neglects the cross second partial derivatives and approximates the values of the homogeneous second partial derivatives by calculating:

$$\frac{\partial^2 f_i}{\partial x_j^2} \approx \frac{\left.\frac{\partial f_i}{\partial x_j}\right|_{x_j + \Delta x_j} - \left.\frac{\partial f_i}{\partial x_j}\right|_{x_j}}{\Delta x_j} \tag{2.24}$$

where Δx_j is the change required in each variable constructional parameter to reduce the value of the merit function. Dilworth called this method the pseudo second derivatives (PSD) method. After some experiments with this method Dilworth in [9] modifies Eq. (2.24) by adding an extra term in the denominator.

$$\frac{\partial^2 f_i}{\partial x_j^2} \approx \frac{\left.\frac{\partial f_i}{\partial x_j}\right|_{x_j + \Delta x_j} - \left.\frac{\partial f_i}{\partial x_j}\right|_{x_j}}{\Delta x_j + \varepsilon} \tag{2.25}$$

where ε is the stabilizing factor to prevent instability in the approximation of the second partial derivatives. If the value of the change of variables Δx_j were to be extremely small at a particular iteration, as sometimes happens, the approximation would give a very large value for the second partial derivative causing the next value for Δx_j to be even smaller. Dilworth used an arbitrary value of 0.001 for the stabilization factor ε. Dilworth in [9] shows that the stabilization factor is necessary because the cross second partial derivatives, which are neglected, have a considerable influence on the approximation.

It is important to notice that the progress in optimization methods has been obtained by increasing the accuracy of the implied assumptions rather than any increase in the actual computation speed or by obtaining additional information about the optical system. So Dilworth in [9] tries to approximate the influence of the cross second partial derivatives. He proposes to deal with the cross second partial derivatives in a statistical sense. He describes the cross second partial derivatives in the following way:

– values of the cross second partial derivatives are randomly distributed;
– the mean value of the cross second partial derivatives is approximately zero;
– the standard deviation σ of the cross second partial derivatives is approximately equal to the value of the homogeneous second partial derivatives.

These assumptions mean that the cross second partial derivatives are expected to be more or less of the same magnitude as the homogeneous second partial derivatives, but with random variations in sign and magnitude. With these assumptions Dilworth in [9] derives a new approximation for the second partial derivatives:

$$\frac{\partial^2 f_i}{\partial x_j^{\,2}} \approx \frac{\left.\dfrac{\partial f_i}{\partial x_j}\right|_{x_j + \Delta x_j} - \left.\dfrac{\partial f_i}{\partial x_j}\right|_{x_j}}{\left|\Delta x_j\right| + \sqrt{\displaystyle\sum_k \Delta x_k^2}\;\Bigg|_{k \neq j}} \tag{2.26}$$

With this approximation the stabilization factor ε is replaced by the term in which the effects of the cross second partial derivatives appear on a statistical basis. It is clear that early in the optimization process, when the variables are undergoing large changes, the quantity under the radical in Eq. (2.26) is large, while at the end of the optimization process when the variables are changing very little that quantity is very small.

Eq. (2.26) includes the absolute value of the change of the variable Δx_j rather than Δx_j itself because there is no sign information in the radical and the approximate second derivative is to be reduced in magnitude when the radical is large.

Dilworth in [10] further developed the idea proposed in [9]. He researched several ideas to even better approximate the cross second partial derivatives. His approximation in [9] was that the cross second partial derivatives are similar to the homogeneous second partial derivatives. Dilworth in [10] poses the questions about which of two variables in the cross second partial derivatives has greater influence and how the influence of each variable in the cross second partial derivatives can be described. Dilworth in [10] proposed to assume that the contribution to the cross second partial derivatives from the variable of different kind (k) is the same as that of the variable of like kind (j) multiplied by the ratio of this quantity, which is called SEC, for these two variables. A very linear variable (on the average) would logically contribute little to the cross second partial derivative, while a non-linear variable would increase the value of the latter. The new improved approximation of the second partial derivatives is:

$$\frac{\partial^2 f_i}{\partial x_j^{\,2}} \approx \frac{\left.\dfrac{\partial f_i}{\partial x_j}\right|_{x_j + \Delta x_j} - \left.\dfrac{\partial f_i}{\partial x_j}\right|_{x_j}}{\left|\Delta x_j\right| + \sqrt{\displaystyle\sum_k \Delta x_k^2 \dfrac{SEC_k}{SEC_j}}\;\Bigg|_{k \neq j}} \tag{2.27}$$

Dilworth concludes in [10] that with this approximation, in the course of optimization, the approximations for the second partial derivatives becomes more and more accurate.

Huber in [11] proposed a new approach to improve the optimization efficiency of the least squares techniques used in the optimization of optical systems. He calls his method the extrapolated least squares (ELS) optimization. The ELS approach to least squares optimization reduces the number of times that the Jacobian matrix ought to be recalculated during the iterative optimization process. This is accomplished by expanding the approximating neighbourhood over which optimization can progress during each optimization iteration. With the ELS method the neighbourhood is defined by a quadratic approximation to the problem. It is well known that classical least squares used linear approximation. It is important to note that within the ELS quadratic neighbourhood linear least squares optimization steps must be taken. However, within this ELS neighbourhood the sequence of least squares steps is made without recalculating the Jacobian matrix.

In order to be started the ELS method needs information (the aberrational functions and the Jacobian matrix) from the last two classical least squares iterations. This information is used to develop the quadratic extrapolation factors. The solution vector ΔX_i determined from the last least squares iteration is used for calculating the ELS extrapolation factors. The optical system, as described by the vector of aberrational functions, is updated at the new location in the parameter space $X_{i+1} = X_i + \Delta X_i$ by using the second order Taylor series approximation for the vector of aberrational functions:

$$f_{i+1} = f_i + \frac{\partial f_i}{\partial x_j} \cdot \Delta X_i + \frac{1}{2} \cdot \frac{\partial^2 f_i}{\partial x_j^2} \cdot \Delta X_i^2 \qquad (2.28)$$

The elements for the Jacobian matrix **A** are given by:

$$\frac{\partial f_{i+1}}{\partial x_j} = \frac{\partial f_i}{\partial x_j} + \frac{\partial^2 f_i}{\partial x_j^2} \cdot \Delta X_i \qquad (2.29)$$

Here $\Delta X_i, f_i, \dfrac{\partial f_i}{\partial x_j}$ have the values retained from the previous least squares iteration step. The second derivative term $\dfrac{\partial^2 f_i}{\partial x_j^2}$ is approximated in the ELS method by:

$$\frac{\partial^2 f_i}{\partial x_j^2} = \frac{\dfrac{\partial f_i}{\partial x_j} - \dfrac{\partial f_{i-1}}{\partial x_j}}{\Delta X_i} \qquad (2.30)$$

where:

$\dfrac{\partial f_i}{\partial x_j}$ is the first partial derivative of the aberration function with respect to j^{th} constructional parameter obtained by the ith least squares iteration;

$\dfrac{\partial f_{i-1}}{\partial x_j}$ is the first partial derivative of the aberration function with respect to j^{th} constructional parameter obtained by the $(i-1)^{th}$ least squares iteration;

ΔX_i is the solution vector obtained by the ith least squares iteration.

The approximation of the second partial derivatives defined in Eq. (2.30) is equivalent to calculating only the elements of the principal diagonal of the Hessian matrix of second partial derivatives. That means that all cross partial derivatives are neglected.

The extrapolated values for the aberrational function and the first partial derivatives, the elements of Jacobian matrix given in equations (2.28) and (2.29) can be used in conjunction with the least squares solution to develop the ELS solution:

$$\Delta \mathbf{X}' = -\left(\mathbf{A}_{i+1}^{\mathbf{T}} \cdot \mathbf{A}_{i+1} + p \cdot \mathbf{I}\right)^{-1} \cdot \mathbf{A}_{i+1}^{\mathbf{T}} \cdot \mathbf{F}_{i+1} \qquad (2.31)$$

where:

$\mathbf{A}_{i+1}$ is the Jacobian matrix calculated by Eq. (2.28);

p is the additive damping factor;

$\mathbf{I}$ is the unit matrix;

$\mathbf{F}_{i+1}$ the vector of aberrational functions calculated by Eq. (2.29).

The solution step $\Delta \mathbf{X}'$ is calculated as an extrapolation of the previous DLS iterative step i.e. this solution step is not recomputed from a new optical system configuration with a new vector of aberrational functions and the Jacobian matrix but simply from updated information provided directly by the ELS procedure.

Matsui and Tanaka in [12] have developed a new analytical method for setting an adequate initial value for the damping factor in the damped least squares optimization by analysing the relationship between eigenvalues of the product of the Jacobian matrix and the damping factor. After a complete mathematical analysis of the damped least squares optimization they conclude that the damping factor should be greater than the minimum of eigenvalues and should be smaller than the maximum eigenvalue. The appropriate value of the damping factor is the median of a series of eigenvalues.

Robb in [13] proposed an alternative way of accelerating convergence in the damped least squares optimization. His approach was based on the observation that the components of the variable constructional parameter vector were often very similar between successive iterations when the DLS optimization was stagnating. An acceleration algorithm was developed that recursively scales the constructional parameter vector after each iteration, increasing the scale factor exponentially until the merit function passes through a minimum.

Robb uses the DLS with the multiplicative damping as a method for optimization. He noticed that the merit function is formed by two independent functions: the optical merit function, which is a sum of squares of the weighted aberrations, and the damping function. A change of constructional parameters Δx_i, which reduces the aberrations i.e. the optical merit function, will always increase the size of the damping function. What happens when the merit function is stagnating is as follows: if the change of constructional parameters Δx_i is large and capable of producing a significant reduction in the optical merit function, it will automatically increase the size of the damping function, and the designer will see only a slight improvement in the whole merit function as a result. The classical way of calculating new constructional parameters is $X_{i+1} = X_i + \Delta x_i$ and Robb proposes a new way of calculating new values for the constructional parameters:

$$X_{i+1} = X_i + \beta^n \cdot \Delta x_i \qquad\qquad (2.32)$$

where β is a suitably chosen scalar quantity. This equation is recursively repeated until n exceeds a specified number of iterations or aberrations for a new optical system are greater than for an old optical system $(f_{i+1} > f_i)$. This approach may be imagined that scaling defined in Eq. (2.32) simply moves the optical system down a multidimensional metric at a faster rate than the program would otherwise take.

2.4   Spencer's optimization

This and following sections will describe alternative optimization methods that were proposed for the optimization of optical systems. The first of them is proposed by Spencer in [14]. He points out that there are two distinct philosophies in the development of the optimization methods for the optical system optimization. The first of these claims that the optimization of optical systems can be completely stated in explicit mathematical terms and, hence, that a computer can reasonably be expected to carry out the entire optimization of optical systems. The second philosophy claims that optimization of optical systems requires qualitative judgments and compromises to be made and, hence, that the computer could be regarded as a tool capable of presenting the designer with provisional solutions only. Spencer thinks that the second philosophy is the most adequate for the optimization of optical systems. So he in [14] formulates the optimization method with sufficient flexibility to allow the designer a high degree of control over the nature of the design solution. Spencer optimization method deals with two kinds of requirements: absolute and relative. Absolute requirements are those requirements which must be met exactly by the design solution. Relative requirements are those which can be lumped together, with various weighting factors, into a single merit function the value of which is to be minimized.

Spencer in [14] proposes a following type of the merit function:

$$\psi = \sum_{k=1}^{K} (\omega_k \cdot (f_k - c_k))^2 \qquad (2.33)$$

where:

ω_k is the weight factor;

f_k is the performance function;

c_k is the constraint representing the desired value for the performance function.

Spencer notices that there will be certain functions f_i for which the specified values are required to be attained, while for the remaining functions the minimization of the merit function will suffice. He divides all performance functions into two groups:

– functions that can be minimized $g_m(x_1, x_2, \ldots, x_J)$, $m = 1(1)M$ where J is the number of the variable constructional parameters.

– functions that must be satisfied $h_n(x_1, x_2, \ldots, x_J)$, $n = 1(1)N < J$.

Then the merit function is minimized:

$$\psi = \sum_{m=1}^{M} (\omega_m \cdot (g_m - c_m))^2 \qquad (2.34)$$

while at the same time securing a simultaneous solution to the set of equations:

$$h_n(x_1, x_2, \ldots, x_J) = t_n \qquad n = 1(1)N < J \qquad (2.35)$$

where the t_n are required values of the functions that must be satisfied.

The main problem in solving equations (2.34) and (2.35) is non-linearity of the functions which measure the optical system performance. A common approach, which is also accepted by Spencer, is to approximate the non-linear functions by linear ones and solve the linear equations.

The starting optical system can be defined by a set of configuration variables $x_1, x_2, \ldots, x_J$. Each of the performance functions may be expanded in a Taylor series about the starting optical system point and neglecting all but the first-order term in the expansion. The following quantities can be defined as:

$$d_m = c_m - g'_m$$
$$e_n = t_n - h'_n$$
$$\Delta x_j = x_j - x'_j$$

$$a_{mj} = \left(\frac{\partial g_m}{\partial x_j}\right)'$$

$$b_{nj} = \left(\frac{\partial h_n}{\partial x_j}\right)'$$

where the primes denote values for the starting optical system. The optimization problem is transformed into the problem of minimizing the quadratic form:

$$\psi = \sum_{m=1}^{M} \omega_m^2 \cdot \left(\sum_{j=1}^{J} a_{mj} \cdot \Delta x_j - d_m\right)^2 \tag{2.36}$$

while at the same time securing a simultaneous solution to the set of linear equations:

$$\sum_{j=1}^{J} b_{nj} \cdot \Delta x_j = e_n \qquad n = 1(1)N \tag{2.37}$$

The problem presented above is a problem of finding an extremum of a function subject to constraints on the variables and that can be solved by using the well-known method of Lagrange multipliers. Spencer in [14] gives a detailed derivation of this method from which only the final equations are stated:

$$\left(\mathbf{M}^T \cdot \mathbf{M} + \mathbf{C}\right) \cdot \Delta \mathbf{X} - \mathbf{B}^T \cdot \lambda = \mathbf{M}^T \cdot \mathbf{R}$$
$$\mathbf{B} \cdot \Delta \mathbf{X} = \mathbf{E} \tag{2.38}$$

where:

$\mathbf{M} = \mathbf{W} \cdot \mathbf{A}$

$\mathbf{W}$ is the diagonal matrix of the weight factors;

$\mathbf{A}$ is the matrix with the first partial derivatives of the performance function that need to be minimized g_m with respect to the constructional parameter x_j;

$\mathbf{C}$ is the diagonal matrix of damping factors which prevents the singularity of the matrix $\mathbf{M}^T\mathbf{M}$ and control the influence of different variables on the solution;

$\Delta \mathbf{X}$ is the vector of constructional parameters changes;

$\mathbf{B}$ is the matrix with the first partial derivatives of the performance function that need to be satisfied h_n with respect to the constructional parameter x_j;

λ is the Lagrange multipliers;

$$\mathbf{R} = \mathbf{W} \cdot \mathbf{D}$$

D is the vector with the differences of the performance function that need to be minimized g_m from the desired value c_m ;

E is the vector with the differences of the performance function that need to be satisfied h_n from the desired value t_n .

2.5   Grey's orthonormal optimization

David S. Grey published two papers in the Journal of the Optical Society of America in 1963 [15] and [16]. He intended to establish a new theory of aberrations but his papers become famous primarily because of the practical success of his computer program in which he used a new orthonormal theory of aberrations in the optimization of optical systems.

Grey in [15] first defines an aberration theory as a classification of image defects according to some useful mathematical formulation. The aberrations are a set of functions which describe the mathematical form of the separate image defects. The aberration depends on a multidimensional variable **Y**. The components of **Y** can be the angular position of the object point and the position of the ray on an entrance or exit pupil. Corresponding to each aberration $f_i(y)$ there is an aberration coefficient $g_i(x)$, which is a function of a multidimensional variable **X** , components of which are the construction parameters of the system. The aberration of a ray defined by the coordinates **Y** when the construction parameters have value **X** is given by:

$$\sum_{i=1}^{N} g_i(x) \cdot f_i(y) \tag{2.39}$$

The classical aberration theory is a development of aberrations according to the Taylor's series with respect to the y parameters. Grey proposes a new orthonormal theory of aberrations. In [15] he first defines an orthogonal aberrational theory which fulfil the following condition:

$$\left(\frac{\partial}{\partial g_i}\right)\left(\frac{\partial \psi}{\partial g_j}\right) = 0 \qquad (i \neq j) \tag{2.40}$$

The classical aberration theory is not orthogonal, hence to some extent residual values of one aberration can be compensated by nonzero residual values of other aberration. If the aberration theory is orthogonal there is no longer the troublesome question of balancing aberration coefficients but the more dominant problem remains: Every parameter affects every aberration coefficient.

The coordinate base of the construction parameters $\mathbf{X}$ can be replaced by the new multidimensional variable $\mathbf{U}$ with components u_i, which are functions of the construction parameters x_i and there are as many u_i as there are x_i. The coordinate base $\mathbf{U}$ is an orthogonal coordinate base if:

$$\frac{\partial g_j}{\partial u_i} = 0 \qquad (i \neq j) \tag{2.41}$$

The aberration theory is orthonormal if it is orthogonal and if:

$$\frac{\partial g_i}{\partial u_i} = 1 \tag{2.42}$$

The specifications for an orthonormal aberration theory are so broad that they cannot be fulfilled precisely. They can be fulfilled approximately, but only to the extent that the second derivatives of the optical aberrations with respect of construction parameters can be ignored. Within any region of the parameter space sufficiently small the second derivatives of the aberrations can be ignored, an orthonormal aberration theory can be constructed. Within this region the transformation from the $\mathbf{X}$ coordinate base to the $\mathbf{U}$ coordinate base can be linear.

Grey uses the classical least squares method, which is defined by Eq. (2.9). These equations are non-linear and ill-conditioned and he tries to deal with both problems by transforming the vector of constructional parameters $\mathbf{X}$ into the orthonormal vector $\mathbf{U}$. The classical mathematical method for this is to premultiply the vector of the constructional parameters $\mathbf{X}$ by a matrix $\mathbf{H}$ whose rows are the eigenvectors of the matrix $\mathbf{A}^T\mathbf{A}$:

$$\mathbf{H} \cdot \mathbf{A}^T \cdot \mathbf{A} \cdot \mathbf{H}^{-1} \cdot \mathbf{H} \cdot \mathbf{X} = -\mathbf{H} \cdot \mathbf{A} \cdot \mathbf{F_0}$$
$$\mathbf{Q} \cdot \mathbf{U} = -\mathbf{V} \tag{2.43}$$

where $\mathbf{Q} = \mathbf{H}\mathbf{A}^T\mathbf{A}\mathbf{H}^{-1}$ is the diagonal matrix, $\mathbf{U} = \mathbf{H}\mathbf{X}$ is the new transformed orthonormal vector and $\mathbf{V} = \mathbf{H}\mathbf{A}^T\mathbf{F_0}$. The least squares equations are transformed in a new set of equations such that each equation now contains only one unknown. The merit function can be minimized with respect to each new variable u_i in turn, knowing that minimization with respect to any one of the variables does not upset any previous minimization with respect to any other variable. The classical method of evaluating the eigenvector matrix $\mathbf{H}$ is a difficult and time-consuming procedure. So Grey *et al.* in [17] proposes the Gramm-Schmidt transformation for the orthonormalization. They have shown that it is always possible to find a triangular matrix which by pre- or postmultiplication will convert an arbitrary matrix into a orthonormal matrix. Applying this to the matrix $\mathbf{A}^T\mathbf{A}$ the following is obtained:

$$\left(\mathbf{A}^T \cdot \mathbf{A}\right) \cdot \mathbf{R} = \mathbf{D} \qquad (2.44)$$

where $\mathbf{R}$ is the upper triangular matrix and $\mathbf{D}^T\mathbf{D} = \mathbf{I}$. Now if the vector of constructional parameters $\mathbf{X}$ is transformed into the orthonormal vector $\mathbf{U}$ such that $\mathbf{X} = \mathbf{R} \cdot \mathbf{U}$ the least squares equations become:

$$
\begin{aligned}
\mathbf{A}^T \cdot \mathbf{A} \cdot \mathbf{X} &= -\mathbf{A}^T \cdot \mathbf{F}_0 \\
\mathbf{A}^T \cdot \mathbf{A} \cdot \mathbf{R} \cdot \mathbf{U} &= -\mathbf{A}^T \cdot \mathbf{F}_0 \\
\mathbf{D} \cdot \mathbf{U} &= -\mathbf{A}^T \cdot \mathbf{F}_0 \\
\mathbf{U} &= -\mathbf{D}^T \cdot \mathbf{A}^T \cdot \mathbf{F}_0
\end{aligned}
\qquad (2.45)
$$

The orthonormal matrix $\mathbf{D}$ and the transformation matrix $\mathbf{R}$ are generated by the Gramm-Schmidt algorithm. First the generation of the orthonormal matrix $\mathbf{D}$ will be described. Let $\mathbf{A}_j$ and $\mathbf{D}_j$ be the jth columns of the matrices $\mathbf{A}^T\mathbf{A}$ and $\mathbf{D}$ respectively. $\mathbf{D}_1$ is simply a scalar multiple of $\mathbf{A}_1$ such that:

$$\mathbf{D}_1^T \cdot \mathbf{D}_1 = \mathbf{I} \qquad \text{and} \qquad \mathbf{D}_1 = \frac{\mathbf{A}_1}{\sqrt{\mathbf{A}_1^T \cdot \mathbf{A}_1}} \qquad (2.46)$$

$\mathbf{D}_2$ is made orthogonal to $\mathbf{D}_1$ by taking $\mathbf{D}_2$ in the direction given by the normalized $\mathbf{A}_2$ vector less than part of the normalized $\mathbf{A}_2$ vector which is parallel to $\mathbf{D}_1$:

$$\frac{\mathbf{A}_2}{\sqrt{\mathbf{A}_2^T \cdot \mathbf{A}_2}} - \frac{\mathbf{A}_2^T \cdot \mathbf{D}_1}{\sqrt{\mathbf{A}_2^T \cdot \mathbf{A}_2}} \cdot \mathbf{D}_1 \qquad (2.47)$$

Normalizing, it is obtained:

$$\mathbf{D}_2 = \frac{\mathbf{A}_2 - \left(\mathbf{A}_2^T \cdot \mathbf{D}_1\right) \cdot \mathbf{D}_1}{\sqrt{\left(\mathbf{A}_2^T \cdot \mathbf{A}_2\right) \cdot \left[\mathbf{I} - \dfrac{\left(\mathbf{A}_2^T \cdot \mathbf{D}_1\right)^2}{\mathbf{A}_2^T \cdot \mathbf{A}_2}\right]}} \qquad (2.48)$$

Continuing in this manner the kth column of $\mathbf{D}$ is:

$$\mathbf{D}_k = \frac{\mathbf{A}_k - \sum_{i=1}^{k-1}\left(\mathbf{A}_k^{\mathrm{T}}\cdot\mathbf{D}_i\right)\cdot\mathbf{D}_i}{\sqrt{\left(\mathbf{A}_k^{\mathrm{T}}\cdot\mathbf{A}_k\right)\cdot\left[\mathbf{I} - \dfrac{\sum_{i=1}^{k-1}\left(\mathbf{A}_k^{\mathrm{T}}\cdot\mathbf{D}_i\right)^2}{\mathbf{A}_k^{\mathrm{T}}\cdot\mathbf{A}_k}\right]}} \tag{2.49}$$

The elements r_{ij} of the coordinate transformation matrix $\mathbf{R}$ are obtained by using the relation $\left(\mathbf{A}^{\mathrm{T}}\cdot\mathbf{A}\right)\cdot\mathbf{R} = \mathbf{D}$:

$$r_{kk} = \sqrt{\left(\mathbf{A}_k^{\mathrm{T}}\cdot\mathbf{A}_k\right)\cdot\left[\mathbf{I} - \dfrac{\sum_{i=1}^{k-1}\left(\mathbf{A}_k^{\mathrm{T}}\cdot\mathbf{D}_i\right)^2}{\mathbf{A}_k^{\mathrm{T}}\cdot\mathbf{A}_k}\right]}$$

$$r_{ik} = -r_{kk}\sum_{j=1}^{k-1}\left(\mathbf{A}_k^{\mathrm{T}}\cdot\mathbf{D}_j\right)\cdot r_{ij} \qquad (i < k)$$

$$r_{ik} = 0 \qquad\qquad\qquad (i > k) \tag{2.50}$$

The optimization with respect to the kth variable may then be summarized as follows:

– compute $\mathbf{D}_k$ from $\mathbf{A}_k$ and all previous columns of $\mathbf{D}$;

– solve the transformed least squares equations, which is now a trivial operation.

The orthonormalization allows a detailed treatment of the difficulties in solving the least squares equations arising from the ill-condition of the equations and from non-linearities.

Ill-condition shows itself as the generation of orthogonal vectors of extremely short length. This means that $\mathbf{A}_k$ used to generate $\mathbf{D}_k$ is not independent. Geometrically this means that the N-dimensional space of the construction parameters is in fact nearly $(N-1)$-dimensional. The use of this vector in optimization will result in unreliable results, and to negate this effect one can set the corresponding element of $\mathbf{U}$ either to zero or to a small predefined value.

Non-linearity results in a lack of correlation between the predicted aberration change and the actual change. If the correlation is poor for a particular orthogonal variable then the predicted step in this variable is reduced. As a last resort it may be set to zero, in which case it will have no effect in construction of further orthogonal variables.

2.6 Simulated Annealing

The ideas forming the basis of simulated annealing were first published by Metropolis *et al.* [18] in 1953 in an algorithm to simulate the cooling of material in a heat bath – a process known as annealing. The annealing essentially is the process of heating solid material beyond the melting point and cooling back into solid state. The annealing process can be simulated by regarding the material as a system of particles. Essentially, the Metropolis's algorithm simulates the change in energy of the system when subjected to a cooling process, until it converges to a steady "frozen" state. Thirty years later, 1983, Kirkpatric *et al.* in [19] suggested that this simulation could be used as an optimization method. This paper published in the Science initiated a great interest and substantial number of researchers tried to apply the simulated annealing into the optimization of various technical systems.

The simulated annealing is based on the laws of statistical mechanics and thermodynamics. Metropolis defined the algorithm that can be used to provide an efficient simulation of a collection of atoms in equilibrium at a given temperature. It is well-known that the atoms in the state of equilibrium possess minimum energy. In each step of this algorithm, an atom is given a small random displacement and the resulting change ΔE in the system energy is computed. If $\Delta E \leq 0$, the displacement is accepted and the configuration with the displaced atom is used as a starting point for the next step. The case $\Delta E > 0$ is treated probabilistically. The probability that the configuration is accepted is defined by the following equation:

$$P(\Delta E) = \exp\left(\frac{-\Delta E}{k \cdot T}\right) \tag{2.51}$$

where:

k is Boltzmann's constant;

T is the current temperature.

Eq. (2.51) defines the famous Boltzmann distribution law. Random numbers uniformly distributed in the interval (0,1) are convenient means of implementing a random part of the algorithm. One such number is selected and compared with the probability $P(\Delta E)$. If the random number is less than the probability $P(\Delta E)$ the new configuration is retained; if not, the original configuration is used to start the next step. By repeating the basic step many times, the thermal motion of atoms at the temperature T is simulated.

In a relatively simple way it is possible to change from the simulation of the thermodynamic processes to optimization. Reeves in [20] gives a table which converts the variables used in thermodynamics into the variables used in the combinatorial optimization.

Table 2.1. Correlation between the thermodynamics and the combinatorial optimization

Thermodynamic simulation	Combinatorial optimization
System states	Feasible solutions
Energy	Merit function
Change of state	Neighbouring solution
Temperature	Control parameter
Frozen state	Heuristic solution

It should be noted that a temperature in the optimization is an ordinary control parameter expressed in same units as the merit function and has no physical meaning.

Schwefel in [21] gives detailed description of the algorithm in pseudocode:

Step 1: (Initialization)

Choose: a start position $x^{(0,0)}$

a start temperature $T^{(0)}$

a start width $d^{(0)}$ for the variations of x

Set $x^* = \tilde{x} = x^{(0,0)}$, $k = 0$, $l = 1$

Step 2: Metropolis simulation

Construct $x^{(k,l)} = \tilde{x} + d^{(k)} \cdot z$

where z is uniformly distributed for all components z_i and for all

$i = 1(1)n$ in the range $z \in \left[-\frac{1}{2}, +\frac{1}{2} \right]$ or normally distributed according to

$$w(z) = \frac{1}{\sqrt{2\pi}} \exp\left(-\frac{1}{2} z_i^2 \right)$$

If $\psi\left(x^{(k,l)}\right) < \psi\left(x^*\right)$, set $x^* = x^{(k,l)}$

If $\psi\left(x^{(k,l)}\right) < \psi(\tilde{x})$ go to step 4

otherwise draw a uniform random number χ from the interval $[0,1]$

If $\chi \le \exp\left(\frac{\psi\left(x^{(k,l)}\right) - \psi(\tilde{x})}{T^{(k)}} \right)$ go to step 4

Step 3: Check for equilibrium

If $\psi(x^*)$ has not been improved within the last N trials, go to step 5

Step 4: Inner loop

Set $\tilde{x} = x^{(k,l)}$ increase $l \leftarrow l + 1$ and go to step 2

Step 5: Termination criterion

If $T^{(k)} \le \varepsilon$ end the search with result x^*

Step 6: Cooling outer loop

Set $x^{(k+1,0)} = x^{*}, \tilde{x} = x^{*}$

and $T^{(k+1)} = \alpha T^{(k)}, \quad 0 < \alpha < 1$

Eventually decrease $d^{(k+1)} = \beta d^{(k)}, \quad 0 < \beta < 1$

Set $l = 1$, increase $k \leftarrow k+1$ and go to step 2

The most important feature of the simulated annealing is the ability to escape from inferior local minimums by allowing deteriorations with a certain probability. This kind of "forgetting principle" is not present in most numerical optimization methods.

Hearn in [22] used simulated annealing in the optimization of optical systems. He used two forms of the probability function. The first function is called the conventional probability function:

$$P(c) = \begin{cases} \exp\left(-\beta \cdot \dfrac{\Delta \psi}{\psi}\right) & za \quad \Delta \psi > 0 \\ 1.0 & za \quad \Delta \psi < 0 \end{cases} \tag{2.52}$$

where:

$P(c)$ is the conventional probability function;

$\Delta \psi$ is the change of the current optical system merit function compared to the previous accepted optical system;

ψ is the last accepted optical system merit function;

β is the scaling factor inversely proportional to the temperature;

The second function is called the natural probability function:

$$P(n) = \frac{1}{1 + \exp\left(-\beta \cdot \dfrac{\Delta \psi}{\psi}\right)} \quad \text{for all values of } \Delta \psi \tag{2.53}$$

where $P(n)$ is the natural probability function.

The natural probability function is derived from the Boltzmann local partition function and has the advantage of yielding a 50% probability of accepting optical system when the magnitude of the change of merit function $\Delta \psi$ is small.

Conversely, the conventional probability function will produce probability greater than 50% when the change of the merit function $\Delta \psi$ is a small positive number.

The result of this is that the likelihood of accepting small deterioration of the optical system merit function is greater than 50%.

The quantity $\frac{\Delta\psi}{\psi}$ is equivalent to a percentage change in the merit function of the current optical system relative to the last accepted optical system. To give the scaling factor β more physical significance Hearn used an $A\Delta\psi(50)$ term which defines the percentage of the merit function increase that will result in the probability of acceptance of 50%. For example if $A\Delta\psi(50) = 2.5\%$ then the value for the scaling factor β is calculated that will result in the 50% probability of the optical system acceptance when a 2.5% increase in the merit function is observed.

The scaling factor β represents a trade-off between the rate of convergence and a tendency to escape local minima. When the scaling factor β is large, the probability of accepting an optical system with a deteriorating merit function falls thus making convergence more likely. The rate of convergence is not very good even with a large value for the scaling factor β because the random walk does not produce an efficient path to the minimum location. If the value for the scaling factor β is small, escape from local minima is easy but convergence is even less efficient.

Hearn, discussing the positive and the negative features of the simulated annealing, concludes that the major practical strength of simulated annealing is ability to escape from local minima and eventually locate a global minimum. He in [22] proposes to combine two optimization methods: simulated annealing and damped least squares. Then the simulated annealing will overcome the major weakness of DLS, getting stuck in a nearest local minimum, while the rapid convergence of DLS will overcome the two weaknesses of the simulated annealing, the poor convergence and the sensitivity to an $A\Delta\psi(50)$ value selection.

2.7   Glatzel's adaptive optimization

All described optimization methods have one thing in common. The number of aberrations were always greater then the number of variable constructional parameters. Glatzel's adaptive optimization method is the first optimization method where the number of variable constructional parameters are greater than the number of the aberrations. The optimization method is described in papers by Glatzel and Wilson [23] and Rayces [24].

An optical system is defined by a set of constructional parameters x_j. The properties of the optical system are characterized by a set of functions $\phi_i(x_j)$ which are generally aberrations but can also be focal length, magnification, image height, etc. When optical system does not meet certain demand, i.e. the value of the function $\phi_i(x_j)$ is greater than expected, a target t_i is set and error function is formed $f_i = t_i - \phi_i$ and reduced to zero by appropriate changes in the constructional parameters, if possible at all.

Glatzel uses a well-known least squares optimization in a form:

$$\mathbf{F} = \mathbf{A} \cdot \Delta \mathbf{X} \tag{2.54}$$

where:

$\mathbf{F}$ is the vector of residual errors i.e. differences between targets and aberrations;

$\mathbf{A}$ is the Jacobian matrix – matrix of first partial derivatives of the ith aberrational function with respect to the jth constructional parameter $a_{ij} = \dfrac{\partial \phi_i}{\partial x_j}$;

$\Delta \mathbf{X}$ is the vector of constructional parameter changes.

When the number of parameters is larger than the number of residual errors to the equations defined by Eq. (2.54) the condition $\| \Delta \mathbf{X} \| = \min$ ought to be added where $\| \Delta \mathbf{X} \|$ is the length of the vector of constructional parameter changes in the Euclidean space. The solution to Eq. (2.54) with this condition is sometimes called the minimal solution:

$$\Delta \mathbf{X} = \mathbf{A}^{\mathrm{T}} \cdot \left(\mathbf{A} \cdot \mathbf{A}^{\mathrm{T}} \right)^{-1} \cdot \mathbf{F} \tag{2.55}$$

The solution obtained by Eq. (2.55) ought to be kept within the linear approximation. This can be done with two methods. The first method is to reduce the length of the vector of constructional parameter changes $\| \Delta \mathbf{X} \|$ by multiplying the right side of Eq. (2.55) by a scalar $q \leq 1$:

$$\Delta \mathbf{X} = q \cdot \mathbf{A}^{\mathrm{T}} \cdot \left(\mathbf{A} \cdot \mathbf{A}^{\mathrm{T}} \right)^{-1} \cdot \mathbf{F} \tag{2.56}$$

Glatzel in [23] calculates the scalar q in such way that the largest constructional parameter change $\max |\Delta x_j|$ is not lager than the step length, a small predefined quantity. Rayces in [24] calculates the scalar q in such way that the length of the vector of constructional parameter changes $\| \Delta \mathbf{X} \|$ is not larger than the step length.

The second method is to keep the vector of residual errors $\mathbf{F}$ under control. All the residual errors are controlled simultaneously and reduced gradually and automatically.

2.8 Constrained optimization

The addition of constraints to the optimization problem considerably increases the complexity of the theory needed for solving this problem. The problem now becomes:

minimize $\quad \psi(x), \quad\quad x$ in set $\mathbf{R}^n$

subject to $\qquad c_i(x) = 0, \qquad i$ in set E $\hfill$ (2.57)

$\qquad\qquad\qquad c_i(x) \geq 0, \qquad i$ in set I

The set of active constraints, denoted the set A, consists of all equality constaints $(c_i(x) = 0)$ from the set E and those inequality constraints $(c_i(x) \geq 0)$ from the set I on their boundaries. The constrained functions $c_i(x)$ may be either linear or nonlinear functions of the variables.

There are two basic approaches to the constrained optimization:

– convert the constrained problem into an unconstrained problem by penalty function method;

– solve a set of equations based upon the necessary conditions for a solution of the constrained problem by quadratic programming methods.

2.8.1   Penalty function method

The appruch used by penalty function method is to convert the constrained optimization problem into an equivalent unconstrained optimization problem. This is achived by minimization of an unconstrained quadratic penalty function $P(x, \omega)$:

$$P(x, \omega) = \psi(x) + \sum_{i \in A} (\omega_i \cdot c_i(x))^2 \qquad (2.58)$$

where:

$\psi(x)$ is the merit function for the unconstrained optimization;

ω_i is the constraint weights;

$c_i(x)$ is the active constrained functions.

Since the merit function, in the case of optical system optimization, is represented by the sum of squares function, the quadratic penalty function approach can be viewed as adding the constraints into the merit function. The constraint weights are chosen to be large enough so the constraints on the solution are solved within some tolerance. The great problem with this method is a right selection of the constraint weights. If the constraint weights are not optimally chosen, the algorithm may fail to find the solution, i.e. the optimal optical system. The constraint weights must be large enough to hold the constraints close to thier targets but small enough to avoid difficulty of minimizing the penalty function. These goals may be inconsistent. In a general the contribution of the weighted constraint should be comparable to the average aberration in the merit function in magnitude. The constraint weights should be adjusted as necessary to increase or decrease the attention paid to the constraints.

Another problem is the practical implementation of inequality constraints. In the optimization of optical systems inequality constraints such as clear apertures, thicknesses and glasses from glass catalogue are very common and it is desirable to have reliable means of obtaining solutions. One sided penalty function called the barrier function have been proposed but can lead to constraints violently oscillating between feasibility and violation on successive iterations. Reliable algorithms for weighted inequality constraints are complex and unwieldy compared to the Lagrange multipliers.

2.8.2   Quadratic programming methods

Algorithms for the constrained optimization can be developed from the consideration of the necessary and sufficient conditions for a constrained optimum point. In the solution, a linear combination of the active constraint gradients is equal and oposite to the merit function gradient. This condition is expressed in the following equations:

$$g_i = -\sum_{i \in A} \lambda_i \cdot b_i(x)$$

$$\lambda_i \geq 0 \qquad\qquad i \text{ in set } \quad I \cap A$$

(2.59)

where:

$g_i = \dfrac{\partial \psi}{\partial x_i}$ is the merit function gradient, i.e. the first partial derivative of merit function with respect to the constructional parameter;

$b_i(x) = \dfrac{\partial c_i}{\partial x_j}$ is the constraint gradient, i.e. the first partial derivative of the active constrain function with respect to the constructional parameter;

λ_i is the Lagrange multiplier.

The magnitude of the Lagrange multipliers is proportional to the sensitivity of the merit function to changes in the constraint targets. The sign of the Lagrange multipliers indicates whether the constraint is tending towards violation or feasibility.

The most rapid progress towards a solution, i.e. the optimal optical system, requires that the optimization problem is minimally constrained. It makes no sense to solve an inequality constraint when the merit function minimum lies within the constraint's feasible region. In general it is not known which inequalities will be active at the solution point. Indeed the set of inequality constraints will often change during the course of optimization. Availability of Lagrange multipliers allow inteligent decisions to be made on which inequality constraints need to be retained in the active set because the sign of the multipliers indicates the direction of the constraint movement. Thus Lagrange multipliers are a powerful tool for determining the minimum number of constraints to retain in the active set while maintaining feasibility. The Lagrange multipliers can be additionally used by the designer to obtain feedback on the impact of the individual constraints on the solution process. The magnitude of the Lagrange multipliers shows how much the merit function would improve if the constraint were relaxed or removed.

$$\lambda_i = \frac{d\psi}{dc_i(x)} \tag{2.60}$$

This fact allows the designer of optical systems to directly trade off the cost of imposing the constraint against minimization of the merit function. The designer can use the magnitude of the multipliers to spot ill-posed problems, i.e. problems where constraints were inadvertently or incorrectly imposed on the problem. In these cases, very large values of the multipliers indicate that the problem definition, i.e. starting optical system with its constraints, is inconsistent and should be re-examined by the designer.

This is a general description of the Lagrange multipliers. Kidger in [3] describes the Lagrange multiplier applied to the optimization of optical systems. Let N be the number of contructional parameters, M be the number of aberrations and L be the number of constraints that ought to be controled by the Lagrange multipliers. The merit function is defined by Eq. (2.11):

$$\psi = \sum_{i=1}^{M} f_i^2 + p^2 \cdot \sum_{j=1}^{N} x_j^2$$

but the merit function is constrained with the following set of equations:

$$\mathbf{C} + \mathbf{B} \cdot \mathbf{X} = 0 \tag{2.61}$$

wher are:

$\mathbf{C}$ is the vector of contraints $c_i(x)$ that ought to be fulfilled;

$\mathbf{B}$ is the square matrix $L \cdot L$ that contains the first partial derivatives of constraints with respect to the constructional parameter $b_{ij} = \dfrac{\partial c_i}{\partial x_j}$.

In order to satisfy Eq. (2.61) the merit function must be in the form:

$$\psi = \sum_{i=1}^{M} f_i^2 + p^2 \cdot \sum_{j=1}^{N} x_j^2 + 2 \cdot \sum_{i=1}^{L} \lambda_i \cdot \left(\sum_{j=1}^{N} b_{ij} \cdot x_j + c_i \right) \tag{2.62}$$

The minimum of the merit function is:

$$\frac{\partial \psi}{\partial x_j} = 2 \cdot \left(\sum_{i=1}^{M} f_i \cdot \frac{\partial f_i}{\partial x_j} + p^2 \cdot x_j + \sum_{i=1}^{L} \lambda_i \cdot b_{ij} \right) = 0 \tag{2.63}$$

or in the matrix form:

$$\mathbf{A}^{\mathrm{T}} \cdot \mathbf{F}_0 + (\mathbf{A}^{\mathrm{T}} \cdot \mathbf{A} + p^2 \cdot \mathbf{I}) \cdot \mathbf{X} + \mathbf{B}^{\mathrm{T}} \cdot \lambda = 0$$
$$\mathbf{B} \cdot \mathbf{X} = \mathbf{C}$$

(2.64)

This forms the $(N + L)$ system of linear equations to be solved. A disadvantage of using Lagrange multipliers is that the matrix of the first partial derivatives is expanded by one row and one column for every constraint, which is controlled in this way. This means that additional storage space and computing time is needed.

Chapter 3

Genetic Algorithms

Genetic algorithms (GA) are adaptive methods which may be used to solve complex search and optimization problems. They are based on the simplified simulation of genetic processes. Over many generations natural populations evolve according to the principles of natural selection and "survival of the fittest" first described by Charles Darwin in a famous book *The Origin of Species* [25]. By mimicking this process, genetic algorithms are able to "develop – evolve" solutions to real world problems. The foundations of genetic algorithms were first laid down rigorously by Holland in [26] and De Jong in [27]. De Jong first applied genetic algorithms in the optimization.

In nature, individuals in a population compete with each other for resources such as food, water and shelter. Also, members of the same species often compete to attract a mate. The most successful individuals in surviving and attracting mates will have relatively larger numbers of offspring.

Genetic algorithms use a direct analogy with natural behaviour. They work with a population of individuals, each representing a possible solution to a given problem. Each individual is assigned a merit function value according to how good a solution to the problem is. The highly fit individuals are given chance to reproduce by cross breeding with other highly fit parents. The least fit members of the population are less likely to get selected for reproduction and so die out.

A whole new population of possible solutions is thus produced by selecting the best individuals from the current generation and mating them to produce a new set of individuals. This new generation contains a higher proportion of the characteristics possessed by the good members of the previous generation. In this way over many generations, good characteristics are spread through out the population, being mixed and exchanged with other good characteristics as they go. By favouring the mating of the more fit individuals, the most promising areas of the optimization space are explored. This leads the population to converge to an optimal solution to the problem.

The power of genetic algorithms comes from the fact that the technique is robust and can deal successfully with a wide range of problem areas, including those difficult for other methods to solve. The genetic algorithms are not guaranteed to find the global optimum solution to a problem, but they are generally good at finding acceptably good solutions to the problem acceptably quickly. Where specialized techniques exist for solving particular problems they are likely to outperform genetic algorithms in both speed of the convergence to the optimal solution and the accuracy of the final result. The main application for genetic algorithms is in difficult areas where no such techniques exist. Even where existing techniques work well, improvements have been made by hybridising them with the genetic algorithms.

3.1   General description of the simple genetic algorithm

This section will describe general properties of the simple genetic algorithm (SGA) as proposed by Goldberg in [28]. The SGA is the genetic algorithm with only the most basic elements that every genetic algorithm must have. Those elements are:

– the population of individuals;
– the selection according to the individual's merit function,
– the crossover to produce a new offspring and the random mutation of a new offspring;

In the SGA individuals are represented by bit strings i.e. the strings of 0s and 1s. Each individual is assigned the merit function which represent how close to the solution every individual is. In the maximization problem the individual closer to the solution has greater merit function, while in the minimization problem the individual closer to the solution has smaller merit function. In the SGA the process of selection of individuals for the reproduction is following: the better (the greater for the maximization or the smaller for the minimization) merit function an individual has, the more times it is likely to be selected to reproduce.

The crossover operator randomly chooses a position in the individual and exchanges the subsequences before and after that position between two individuals to create two new offspring. For example, the strings 10000100 and 11111111 could be crossed over after the third position in each to produce the two offspring 10011111 and 11100100.

The mutation operator randomly flips some of the bits in the individual. For example, the string 00000100 might be mutated in its second position to become 01000100. The mutation can occur at each bit position in a string with some probability, usually very small (e.g. 0.001).

The algorithm for the SGA is following:

Step 1:   Start with randomly generated population with n l-bit individuals.

Step 2:   Calculate the merit function of each individual in the population.

Step 3:   Repeat steps 4-6 until n offspring have been created:

Step 4:   Select a pair of parent individuals from the current population. The probability of selection is directly proportional to the individual's merit function meaning that the individual with the better merit function has more chance to become a parent more than once.

Step 5: With the crossover probability p_c cross over the chosen pair of parent individuals at a randomly chosen point to form two offspring. If no crossover takes place, form two offspring that are exact copies of their respective parents.

Step 6: Mutate the two offspring at each position with the mutation probability p_m and place the resulting individuals in the new population.

Step 7: Replace the current population with the new population.

Step 8: Repeat steps 2-7 until the optimal solution is found or the computational resources are exhausted.

Each iteration of this process is called a generation. The SGA is typically iterated for anywhere from 50 to 500 or more generations. The entire set of generations is called a run.

The simple procedure just described is a basis for most variations of genetic algorithms. In the next sections the following parts of genetic algorithms will be analysed:

– the various ways for the representation of individuals;
– the different methods for the selection of individuals for the reproduction;
– the numerous genetic operators.

3.2 Representation of individuals in the genetic algorithm

The way in which individuals are represented is a central factor in the success of a genetic algorithm. Most GA applications use fixed length, fixed order bit strings to encode individuals. Some people in the genetic algorithms field have come to believe that the bit strings should be used as encoding technique whenever they apply a genetic algorithm. However, in recent years, there have been experiments with other kinds of representation of individuals in the genetic algorithms like the floating point numbers.

3.2.1 Representation of individuals with the binary numbers

The representation of individuals with the binary numbers use the bit strings (string of binary digits 1 or 0) for encoding of possible solutions to the problem. The optimization of optical system from the mathematical point of view can be considered as a minimization problem of the merit function. The merit function definition is the same as in the classical optimization which is sum of the squares of weighted aberrations. This merit function uses floating point numbers which must be encoded to the bit strings. In order to do it, it is necessary to do the following:

– define the interval in which the possible problem solutions exist;
– define the necessary precision in the floating point numbers representation;

- in accordance with the necessary precision and the interval for floating point numbers the length of bit string necessary to encode the floating point number can be defined;
- the optimization problem is usually defined by several variable constructional parameters expressed by the floating point numbers. All these floating point numbers can be encoded into bit strings and these bit strings concatenated together in a large bit string which consists of all variable constructional parameters.

The bit strings have several advantages over other ways of representing individuals. They are simple to create and manipulate. Their simplicity makes easy to prove theorems about them. Holland in [26] proved the power of natural selection on the bit string encoding. The most theoretical and practical researches have been conducted with the bit strings as a principal way of representing individuals as possible solutions to the problem.

The bit strings have also shortcomings especially when are applied to the multidimensional numerical optimization problems requiring high numerical precision. The formula for calculating the numerical precision is:

$$T = \frac{UB - LB}{2^n - 1} \tag{3.1}$$

where:

T is the numerical precision i.e. the number of exact digits;

UB, LB are the upper and the lower bound of the interval in which the possible solutions exist;

n is the number of bits per one floating point number

Michalewicz in [29] has conducted detailed research in order to find answer to the question whether it is better to use the bit strings or the floating point numbers in the representation of the individuals as a possible solutions to the problem. He constructed two implementations of the genetic algorithm, which were equivalent except they had different ways of encoding individuals. After conducting several experiments Michalewicz has concluded that the floating point representation is faster, more consistent from run to run and provides a higher precision (especially with large domains where bit strings would require prohibitively long representations). In addition, the floating point representation, as intuitively closer to the problem space, is easier for designing other operators incorporating problem specific knowledge. This is especially essential in handling nontrivial, problem specific constraints.

3.2.2   Representation of individuals with the floating point numbers

The representation of individuals with the floating point numbers use the floating point data records, which describe possible solutions to the problem in more details. The representation of individuals with the floating point numbers looks more natural because they can use description of the possible solution to the problem from the existing classical optimization methods. When the individuals are represented with the bit strings the problem is treated as a black box and it is tried to be solved with as few specific knowledge as possible, that is the algorithm tries to be as much general as it is possible. On the other hand, when the individuals are represented with the floating point numbers, then it is tried to include as much specific knowledge as possible to facilitate the problem solving, but the algorithm become highly specialized for solving only one class of the problems, for example the optical systems optimization.

The optical systems optimization is the specific field of optimization with the large specific knowledge gained during the decades of the successful research.

In the optical system optimization by the genetic algorithm the representation of individuals with the floating point numbers is used.

3.3   Merit function scaling

In all minimization problems the genetic algorithm needs a scaling function which will translate the original merit function values into the positive floating point numbers because the standard selection mechanisms in the genetic algorithms requires that the best individuals have the greatest positive merit function values. In all minimization problems, the problem with the merit function values exists because the better merit function the smaller merit function. That means a better individual with a smaller merit function gets proportionally less chances for the reproduction during selection than a worse individual with a greater merit function. This is in contrast with the theory of genetic algorithms applied in the optimization. The optical systems optimization is the classical minimization problem. In the optical systems optimization the merit function is formed as a sum of squares of weighted aberrations and has always-positive values. It is well known fact that the aberrations are smaller for a better optical system.

Besides just described need for the merit function scaling, Goldberg in [28] describes some other benefits of the merit function scaling. The main advantage is keeping appropriate levels of competition throughout the genetic algorithm. Without the merit function scaling, in the early phase of the optimization there is a tendency for a few super individuals to dominate the selection process. In this case the merit functions must be scaled back to prevent takeover of the population by these super individuals. Later on when the population is largely converged, competition among the population members is less strong and the genetic algorithm optimization tends to wander. In this case the merit functions must be scaled up to accentuate differences between the population members to continue to reward the best performers.

3.3.1   Linear scaling

The linear scaling is described in [28] and [30]. The new, scaled, value for the merit function (f') is calculated by linear scaling when the starting value for the merit function (f) is inserted in the following linear equation:

$$\psi' = a \cdot \psi + b \qquad (3.2)$$

where the coefficients a and b are usually chosen to satisfy two conditions:
- the first is that the average scaled merit function is equal to the average starting merit function;
- the second is that the maximum-scaled merit function is a specified multiple (usually two) times greater than the average scaled merit function.

These two conditions ensure that the average population members create one offspring on average and the best create the specified multiple number of offspring. One possible problem with the linear scaling is a potential negative values appearance for the scaled merit function.

3.3.2   Linear dynamic scaling

Grefenstette and Baker in [30] and Bäck in [31] described the linear dynamic scaling. The equation for the linear dynamic scaling is different for the minimization and the maximization problems.

For the maximization problem the equation is:

$$\psi' = a \cdot \psi - \min(\psi) \qquad (3.3)$$

where $\min(\psi)$ represents the merit function for the worst individual in the population and a represents the coefficient defined in the section 3.3.1.

For the minimization problem the equation is:

$$\psi' = -\psi + \max(\psi, \omega) \qquad (3.4)$$

where $\max(\psi, \omega)$ represents the merit function for the worst individual in the ω generations. Grefenstette in [30] suggests from experience to set $\omega = 5$ as a reasonable default value. The parameter ω is called the scaling window.

3.3.3   Sigma (σ) scaling

The sigma scaling got its name after the sign for the standard deviation (σ). Several researchers proposed different ways for the sigma scaling. The sigma scaling was designed as an improvement of linear scaling. It deals with the negative merit function values and incorporates the problem dependent information into the scaling function. The idea behind the sigma scaling is to keep the selection pressure (i.e. the degree to which highly fit individuals are allowed to have many offspring) relatively constant over the course of the optimization rather than depending on the merit function variances in the population. Under the sigma scaling, an individual's scaled value for the merit function is a function of:

— its merit function;
— the mean and the standard deviation of all merit functions of the population in the current generation.

The first to come with the proposition for the sigma scaling is Forest in 1985 in an unpublished document. His idea of sigma scaling definition is taken from Bäck in [31]:

$$\psi' = \frac{\psi - \left(\overline{\psi} - \sigma\right)}{\sigma} \tag{3.5}$$

where:

$\overline{\psi}$ is the mean value of all merit functions from the individuals in the population for the current generation;

σ is the standard deviation of all merit functions from the individuals in the population for the current generation.

Goldberg in [28] reformulates the sigma scaling in a form:

$$\psi' = \psi - \left(\overline{\psi} - c \cdot \sigma\right) \tag{3.6}$$

where the constant c is chosen as a reasonable multiple of the merit function standard deviation for the population (usually between 1 and 3). The negative merit function values are arbitrary set to 0.

Michalewicz in [29] substitutes the first minus sign in Eq. (3.6) with the plus sign:

$$\psi' = \psi + \left(\overline{\psi} - c \cdot \sigma\right) \tag{3.7}$$

He defines the constant c as an integer value from 1 to 5.

Brill *et al.* in [32] defined the sigma scaling as:

$$\psi' = \frac{\psi - \left(\overline{\psi} - c \cdot \sigma\right)}{2 \cdot c \cdot \sigma} \tag{3.8}$$

Tanese in [33] defined the sigma scaling as:

$$\psi' = \begin{cases} 1 + \dfrac{\psi - \overline{\psi}}{2 \cdot \sigma} & \text{if } \sigma \neq 0 \\ 1.0 & \text{if } \sigma = 0 \end{cases} \tag{3.9}$$

This scaling algorithm gives an individual with the merit function one standard deviation above the mean merit function 1.5 expected offspring. If the scaled merit function is negative then the arbitrary small value (0.1) is assigned to the individual, so that the individuals with very small-scaled merit function had a small chance of reproducing.

3.3.4 Exponential Scaling

The exponential scaling is the scaling method where the starting merit function is taken to some specific power near one. Goldberg in [28] and Michalewicz in [29] suggest value $k = 1.005$:

$$\psi' = \psi^k \tag{3.10}$$

Bäck in [31] suggests the following form of the exponential scaling:

$$\psi' = \left(a \cdot \psi + b\right)^k \tag{3.11}$$

where a and b are the same constants as in the linear scaling defined in the section 3.3.1.

3.3.5 Logarithmic scaling

The logarithmic scaling is the scaling method where the new-scaled merit function is proportional to the logarithm of the starting merit function. This scaling method is proposed in Grefenstette and Baker [30] and Bäck [31].

$$\psi' = c - \log(\psi) \tag{3.12}$$

where the value of the constant c is chosen to satisfy the condition $c > \log(\psi)$.

3.3.6 Linear normalization

Davis in [34] describes the scaling method that can be used in both the minimization problems and the maximization problems. The algorithm for the linear normalization consists of the following steps:

Step 1: Order the individuals of the population so that the best individual is the first and the worst individual is the last one. If the optimization problem is the minimization problem then the best individual has the smallest merit function value.

Step 2: The individual with the best merit function is assigned a constant predefined value (the staring value for the linear normalization).

Step 3: The next individual is assigned a value that is equal to the value of the previous individual decreased linearly with the linear normalization step.

The starting value and the linear normalization step are parameters of the algorithm for linear normalization and can be specified.

The linear normalization will be described in the following example of the minimization problem:

The original merit function	1210	120	30	560	7310
The sorted merit function	30	120	560	1210	7310
The linearly normalized merit function	100	80	60	40	20

The linear normalization is the algorithm that is selected for the scaling of the merit function in the optical systems optimization by the genetic algorithm.

3.4 Selection methods

After deciding on a representation of individuals, the second important decision in using a genetic algorithm is the choice of selection method. The selection method describes how to choose the individuals in the population that will create offspring for the next generation and how many offspring each selected individual will create. The purpose of selection is to enable individuals with the better merit function to create offspring with the even better merit function. The selection has to be balanced with the variation introduced by the crossover and the mutation. Too strong selection means that the sub optimal highly fit individuals will take over the population, reducing the diversity needed for further change and progress of the whole population to the optimal solution. Too week selection will result in too-slow evolution.

Numerous selection methods have been proposed in the GA literature and only some of the most common methods will be described. It is still an open question for genetic algorithm community to clearly define which selection method should be used for which problem.

3.4.1   Merit function proportional selection

One of the first selection methods to be described is the merit function proportional selection. It was described by Holland in [26]. The essence of the merit function proportional selection is the calculation of "expected value" of an individual (i.e. the expected number of times an individual will be selected for reproduction):

$$\eta = \mu \cdot \frac{\psi}{\sum\limits_{i=1}^{\mu} \psi_i} \tag{3.13}$$

where:

η is the expected value of an individual;

μ is the number of individuals in the population;

ψ is the merit function of the current individual;

$\sum\limits_{i=1}^{\mu} \psi_i$ is the sum of the all merit functions from the individuals in the population for the current generation.

The selection method defined by Eq. (3.13) is not appropriate for all optimization problems. If the original, not scaled, merit functions is used then the merit function proportional selection is not appropriate for all minimization problems and those maximization problems where the merit function can be negative. Because of that it is of great importance to use appropriate method for the scaling merit functions.

Two most common ways of the merit function proportional selection implementation are:

– the roulette wheel selection;

– the stochastic universal sampling.

3.4.1.1 Roulette wheel selection

Davis in [34] gives detailed description of the roulette wheel selection. The roulette wheel selection can be described by the following algorithm:

Step 1: Sum the merit functions of all individuals in the population. Call the result the total merit function.

Step 2: Generate a random number between zero and the total merit function.

Step 3: Return the first individual from the population whose merit function added to the merit functions of the preceding individuals from the population is greater than or equal to the generated random number.

The effect of the roulette wheel parent selection is to return a randomly selected parent. Although this selection procedure is random, each parent's choice of being selected is directly proportional to its merit function. On balance, over a number of generations the roulette wheel selection will drive out the individuals with the bad merit functions and contribute to the spread of the individuals with the good merit functions. Of course, it is possible that the individual with the worst merit function in the population could be selected by this algorithm, but the possibility of this event is very small.

The algorithm is referred to as a roulette wheel selection because it can be viewed as allocating pie shaped slices on a roulette wheel to the individuals from the population, with each slice proportional to the individual's merit function. Selection of a population member to be a parent can be viewed as a spin of the wheel, with the winning individual being the one in whose slice the roulette spinner ends up.

3.4.1.2 Stochastic universal sampling

The roulette wheel selection has one shortcoming when the populations are relatively small, the actual number of offspring allocated to an individual is often far from its expected value. Baker in [35] analysed several different algorithms for realizing the spinning wheel with respect to their:

– accuracy measured in terms of bias i.e. deviation between expected values and algorithmic sampling frequencies;

– precision measured in terms of spread i.e. the range of possible values for the number of copies an individual receives by the selection mechanism;

– computational complexity of the sampling algorithm.

Baker in [35] developed an optimal sampling algorithm called the stochastic universal sampling that combines zero bias and minimum spread. The algorithm can be described in the following steps:

Step 1: Generate a random number between zero and one.

Repeat steps 2-4 for each individual in the population.

Step 2: Calculate the expected value of an individual with Eq. (3.13).

Step 3: Calculate the total expected value of an individual. For the first individual in the population the total expected value is equal to its expected value. For all other individuals in the population the total expected value of an individual is equal to the sum of its expected value and the sum of the expected values of all preceding individuals.

Step 4: Compare the total expected value of an individual with the generated random number. If the total expected value is greater, then select that individual to be the parent for the next generation and increase the random number for one. Step four is repeated until the total expected value of an individual is greater than the generated random number.

The stochastic universal sampling can be visualized as spinning the roulette wheel once with N equally spaced pointers, which are used to select N parents. Although the stochastic universal sampling represents improvement in the merit function proportional selection, it does not solve the major problems with this selection method. Typically, early in the optimization the merit function variance in the population is high and a small number of individuals are much fitter than the others. Under the merit function proportional selection they and their descendents will multiply quickly in the population, in effect preventing the genetic algorithm from doing any further exploration. This is known as the premature convergence. In other words the merit function proportional selection early on often puts too much emphasis on exploitation of highly fit individuals at the expense of the exploration of other optimization space regions. Later in the optimization, when all individuals in the population are very similar (the merit function variance is low), there are no real differences in the merit function for the selection to exploit, and the evolution is nearly stopped. Thus, the rate of evolution depends on the merit function variance in the population.

3.4.2   Elitist selection

The elitist selection is first introduced by De Jong in [27]. It represents an addition to many selection methods that forces the genetic algorithm to retain some number of the best individuals at each generation. Such individuals can be lost if they are not selected to reproduce or if they are destroyed by the crossover or the mutation. Many researchers added the elitist selection to their existing methods of selection and they have found that elitism significantly improves the genetic algorithms performance.

3.4.3   Boltzmann selection

It is shown that during the optimization, the different amounts of selection pressure are often needed at different times. At the beginning of the optimization, selection pressure ought to be small i.e. allowing the less fit individuals to reproduce with the probabilities approximately same with the better individuals. Then the selection occurs slowly while maintaining a lot of variation in the population. At the end of the optimization the selection pressure ought to be stronger in order to strongly emphasize better individuals, assuming that the early diversity with slow selection has allowed the population to find the right part of the optimization space.

Goldberg in [36] and de la Maza and Tidor in [37] described Boltzmann selection as an approach similar to the simulated annealing, in which a continuously varying "temperature" controls the rate of selection according to the preset schedule. The optimization starts with the high temperature, which means that selection pressure is low and that every individual has some reasonable probability of reproducing. The temperature is gradually lowered, which gradually increases the selection pressure, thereby allowing the GA optimization to narrow the optimization space while maintaining the appropriate degree of diversity. In a typical implementation each individual is assigned an expected value:

$$\eta = \frac{e^{\frac{\psi}{T}}}{\frac{1}{\mu}\sum_{i=1}^{\mu} e^{\frac{\psi}{T}}} \qquad (3.14)$$

where:

T is the temperature;

μ is the number of individuals in the population;

ψ is the merit function of the current individual.

3.4.4 Rank selection

Rank selection is an alternative selection method whose purpose is also to prevent too-quick convergence. In the rank selection the individuals from the population are ranked according to their merit functions and the expected value of each individual depends on its rank rather than its absolute value of merit function. There is no need to scale the merit function in this case, since absolute differences in merit functions are obscured. This discarding information about the merit function absolute value can have advantages (using information about the merit function absolute value can lead to convergence problems) and disadvantages (in some cases it might be important to know that one individual has far better merit function than its nearest competitor). Ranking avoids giving the far largest share of offspring to small group of individuals with the very good merit functions, and thus reduces the selection pressure when the merit function variance is high. It also keeps up the selection pressure when the merit function variance is low. The ratio of expected values of individuals ranked i and $i+1$ will be the same whether their absolute differences in the merit function are high or low.

Baker in [38] proposed a rank selection method called linear ranking because he used the linear function to approximate the individual expected value. The selection method description is following:

If the optimization problem is the maximization problem then each individual from the population is ranked in increasing order of the merit functions from 1 (the worst individual) to μ (the best individual). The population has μ individuals. The parameter of the algorithm is the expected value for the best individual in the population, η_{max}. The expected value of each individual i in the population is given by:

$$\eta_i = \eta_{min} + (\eta_{max} - \eta_{min})\frac{i-1}{\mu-1} \tag{3.15}$$

where η_{min} is the expected value for the worst individual in the population.

The constraints for this algorithm are:

$$\sum_{i=1}^{\mu} \eta_i = \mu \qquad \text{and} \qquad \eta_i > 0 \quad \forall i \in \{1,...,\mu\}$$

$$1 \le \eta_{max} \le 2 \tag{3.16}$$

$$\eta_{min} = 2 - \eta_{max}$$

The linear ranking selective pressure can be varied by tuning the maximum expected value η_{max}, which controls the linear function slope. Baker in [38] recomends the following value for the maximum expected value $\eta_{max} = 1.1$. This means that the best individual is expected to be selected 1.1 times within the population. This is a rather moderate selective pressure, close to the random optimization i.e. no selective pressure $(\eta_{max} = 1 \Rightarrow \eta_i = 1)$. Once when the expected value for each individual is calculated, the stochastic universal sampling can be used to select individuals in the population for parents.

If the optimization problem is the minimization problem then each individual from the population is ranked in increasing order of the merit functions from 1 (the best individual) to μ (the worst individual). The expected value of each individual i in the population is now given by:

$$\eta_i = \eta_{max} - (\eta_{max} - \eta_{min}) \frac{i-1}{\mu-1} \tag{3.17}$$

with constraints defined by Eq. (3.16).

Michalewicz in [29] has used the exponential ranking method where the expected values are calculated with the following equation:

$$\eta_i = \mu \cdot c \cdot (1-c)^{i-1} \tag{3.18}$$

where $c \, (0 < c \ll 1, \text{ e.g. } c = 0.004)$ is a constant determining the best individual's expected value $\eta_{max} = \mu \cdot c$. Such a method increases selective pressure by emphasizing the best individuals even more. However, it is not widely used and does not fulfil the conditions defined in Eq. (3.16) since:

$$\sum_{i=1}^{\mu} \eta_i = \mu - (\mu - c)^{\mu} < \mu \tag{3.19}$$

3.4.5 Tournament selection

Goldberg and Deb in [39] described the tournament selection. The general definition of the selection method is following: the q-tournament selection method selects a single best individual from the group of q individuals randomly from the population. This selected individual becomes a parent who will produce offspring in the next generation. The tournament is held μ times to fill new population in the next generation. A common tournament size is $q = 2$, also known as binary tournament.

The algorithm for the binary tournament selection is:

Step 1: Select two individuals at random from the population.

Step 2: Generate a random number r between zero and one.

Step 3: If $r < k$ (where k is a parameter, e.g. $k = 0.75$) the individual with the better merit function is selected to be parent, otherwise the individual with the worse merit function is selected. The two selected individuals are then returned to the original population and can be selected again.

The expected value of each individual i in the population for the q-tournament selection is given by:

$$\eta_i = \frac{1}{\mu^q} \cdot \left[(\mu - i + 1)^q - (\mu - i)^q \right]$$

(3.20)

The tournament selection is similar to the rank selection, but it always imposes stronger selective pressure than the rank selection. The tournament selection is computationally more efficient than the rank selection because the rank selection requires sorting the entire population by rank, a potentially time consuming procedure.

3.4.6 Steady-state selection

Most genetic algorithms described in the literature have been "generational", at each generation the new population consists entirely of offspring formed by parents in the previous generation, though some of those offspring may be identical to their parents. In some selection methods such as the elitist selection, successive generations overlap to some degree i.e. the best individuals from the previous generation are retained in the new population. The fraction of new individuals at each generation De Jong in [27] called a generation gap. In the steady-state selection, only a few individuals are replaced in each generation, usually the small number of the individuals with the worst merit function are replaced in each generation with offspring resulting from the crossover and the mutation of the individuals with the best merit function.

Davis in [34] gives an algorithm for the steady-state selection:

Step 1: Create n offspring through the reproduction (the crossover and the mutation).

Step 2: Delete *n* individuals with the worst merit function from the population to make space for the offspring.

Step 3: Evaluate and insert the individuals into the population.

The steady-state selection has one input parameter – the number of new offspring to create. It is typical to create and insert just one or two offspring at a time.

3.4.7   Steady-state selection without duplicates

The steady-state selection without duplicates is the modification of the standard steady-state selection proposed by Whitley in [40] and Syswerda in [41]. It represents a selection technique that discards the offspring that are duplicates of current individuals from the population rather than inserting them into the population. When this selection method is used, every individual in the population is different. The steady-state selection without duplicates has one great benefit, because it allows much more efficient use of new generated offspring. The genetic algorithms that did mot used this selection technique generated a large number of offspring that were exact copies of the individuals from the population.

The steady-state selection without duplicates has been selected as a basic selection method for the optical systems optimization by the genetic algorithm.

3.5   Genetic operators

The third decision to make in implementing a genetic algorithm is what genetic operators to use. This decision depends greatly on the representation of individuals in the population. In the GA literature, the great number of genetic operators is described, which can be divided in two broad groups:

– the crossover operators;

– the mutation operators.

In the GA theory it is considered that the crossover operator is the most important genetic operator. The idea that lies behind the crossover is to recombine the useful parts of different individuals (parents) in order to produce new offspring, which will incorporate the good parts from both individuals. The crossover operator will, during the larger number of generations, form constantly better and better individuals, which will eventually lead to the individual with the optimum merit function.

The mutation is considered in the GA theory as the background operator, which is responsible to keeping diversity of individual in the population. The crossover, as a main genetic operator, researches the optimization space for the individuals that represent good solutions to the optimization problem, while the mutation provides that the population is always diverse and that the crossover has enough good individuals.

3.5.1   Single point crossover

The single point crossover is the basic genetic operator, which appears in almost all genetic algorithm implementations. Holland in [26] defined the single point crossover. The algorithm for the single point crossover is following:

Step 1:   Select two individuals to be parents. Use the appropriate selection method.

Step 2:   Randomly choose the position where the crossing over will happen.

Step 3:   Form two offspring: the first offspring takes the first part of the parent 1 and the second part of the parent 2, while the second offspring takes the second part of the parent 1 and the first part of the parent 2.

Michalewicz in [29] stated the single point crossover algorithm in the mathematical form:

Let the selected parents in the generation t be: $S_v^t = \langle V_1,...,V_k,...,V_m \rangle$ and $S_w^t = \langle W_1,...,W_k,...,W_m \rangle$. If the crossing over happens at the kth component then the resulting offspring will have the following form: $S_v^{t+1} = \langle V_1, ... ,V_k,W_{k+1}, ... ,W_m \rangle$ and $S_w^{t+1} = \langle W_1, ... ,W_k,V_{k+1}, ... ,V_m \rangle$. The offspring have index $t+1$ because they belong to the next generation.

Important caracteristic of all crossover operators is that they can produce offspring that is radically different from its parents. It ought to be noticed that two indentical parents can not produce different offspring.

Bäck in [31] after detailed discussion about crossover operators concludes that the single point crossover in the parameter optimization problems is clearly inferior to other crossover operators with respect to the performance results.

3.5.2   Multiple point crossover

The multiple point crossover is the natural improvement of the single point crossover. It tries to correct all shortcomings of the single point crossover. The most frequent implementation of the multiple point crossover is the two point crossover.

The general algorithm for the n point crossover is not very much different from the single point crossover:

Step 1:   Select two individuals to be parents. Use the appropriate selection method.

Step 2:   Randomly choose the n positions where the crossing over will happen.

Step 3:   Form two offspring by exchanging the segments between the crossover positions.

Michalewicz in [29] stated the two-point crossover algorithm, as an example of the multiple point crossover in the mathematical form:

Let the selected parents in the generation t be: $S_v^t = \langle V_1,...,V_k,...,V_m \rangle$ and $S_w^t = \langle W_1,...,W_k,...,W_m \rangle$. If the crossing over happens at the kth and l^{th} component then resulting offspring will have the following form:

$$S_v^{t+1} = \left\langle V_1, \dots, V_k, W_{k+1}, \dots, W_l, V_{l+1}, \dots, V_m \right\rangle,$$

$$S_w^{t+1} = \left\langle W_1, \dots, W_k, V_{k+1}, \dots, V_l, W_{l+1}, \dots, W_m \right\rangle.$$

The offspring have index $t+1$ because they belong to the next generation.

3.5.3   Uniform crossover

The uniform crossover represents further optimization of the crossover genetic operator and it is proposed by Syswerda in [41]. In case of uniform crossover the process of the crossing over takes place for each position in the individual. This means that for each position in the first offspring it is decided with certain probability p which parent will contribute to its value. The second offspring at that position would receive the value from the other parent. This is the most general way of performing the crossover. The algorithm for the uniform crossover is following:

Step 1:   Select two individuals to be parents. Use the appropriate selection method.

Step 2:   Form two offspring where for each position in the offspring it is decided with certain probability p which parent will contribute to its value.

Michalewicz in [29] stated the uniform crossover algorithm in the mathematical form:

Let the selected parents in the generation t be: $S_v^t = \left\langle V_1, \dots, V_m \right\rangle$ and $S_w^t = \left\langle W_1, \dots, W_m \right\rangle$. If the crossing over happens at the each component then the resulting offspring will have the following form that is a function of randomly choosen components of the parents: $S_v^{t+1} = \left\langle V_1, W_2, W_3, V_4, V_5, \dots, W_m \right\rangle$ and $S_w^{t+1} = \left\langle W_1, V_2, V_3, W_4, W_5, \dots, V_m \right\rangle$. The offspring have index $t+1$ because they belong to the next generation.

The uniform crossover has two good properties:

–   the crossover point is not important;

–   the uniform crossover can combine good propreties wherever they are positioned and however they are distributed. The single or two point crossover represents more local genetic operator because it tries to conserve good properties which are compactly coded.

All three decribed crossover operators are general genetic operators that can be used for all types of representation of individuals in the population (either bit strings or floating point numbers).

In the optical systems optimization by the genetic algorithm the uniform crossover is used as the most general crossover operator. It is possible to imagine some specific types of crossover operators since the floating point numbers are used for the representation of individuals in the optical systems optimization by the genetic algorithm.

3.5.4 Specific crossover operators

Several specific types of crossover operators can be described when the individuals are represented by the floating point numbers. Davis in [34] described the average crossover. Michalewicz in [29] described several types of genetic algorithms in the optimization of various types of problems. System GENOCOP (**GE**netic algorithm for **N**umerical **O**ptimization **CO**nstrained **P**roblems) is the most similar to the problem of the optical system optimization. The following crossover operators are defined:

- the simple crossover;
- the single arithmetical crossover;
- the whole arithmetical crossover;
- the heuristic crossover.

3.5.4.1 Average crossover

Davis in [34] described the average crossover as a crossover that takes two parents and forms one offspring that is the result of averaging the corresponding fields of two parents. The algorithm for the average crossover is following:

Step 1: Select two individuals to be parents. Use the appropriate selection method.

Step 2: Form one offspring where each position in the offspring is the arithmetical mean value of the corresponding components from the parents.

The mathematical definition of the algorithm is following:

Let the selected parents in the generation t be: $S_v^t = \langle V_1, ..., V_m \rangle$ and $S_w^t = \langle W_1, ..., W_m \rangle$. Each component of the reulting offspring will be the arithmetical mean value of the corresponding components from the parents:

$$s_v^{t+1} = \left\langle \frac{V_1 + W_1}{2}, \frac{V_2 + W_2}{2}, ..., \frac{V_m + W_m}{2} \right\rangle \tag{3.21}$$

The offspring have index $t+1$ because they belong to the next generation.

The average crossover operator, together with the uniform crossover operator is used in the optical systems optimization by the genetic algorithm.

3.5.4.2 Simple crossover

Michalewicz in [29] defined the simple crossover to be similar to the single point crossover. There are two important differences. The first is that it uses individuals constructed with the floating point numbers and the second is that it uses the following mathematical property for offspring to stay within the convex solution space S:

For any two points s_1 and s_2 in the solution space S, the linear combination $a \cdot s_1 + (1-a) \cdot s_2$ where $a \in [0,1]$, is the point in the solution space S.

Let the selected parents in the generation t be: $S_v^t = \langle V_1,...,V_m \rangle$ and $S_w^t = \langle W_1,...,W_m \rangle$. If the crossing over happens after the kth component then the resulting offspring will have the following form:

$$S_v^{t+1} = \langle v_1,...,v_k, w_{k+1} \cdot a + v_{k+1} \cdot (1-a),...,w_m \cdot a + v_m \cdot (1-a) \rangle \in S$$
$$S_w^{t+1} = \langle w_1,...,w_k, v_{k+1} \cdot a + w_{k+1} \cdot (1-a),...,v_m \cdot a + w_m \cdot (1-a) \rangle \in S \qquad (3.22)$$

Michalewicz in [29] uses binary search to find the largest value for the coefficient a to obtain the greatest possible information exchange.

3.5.4.3 Single arithmetical crossover

Michalewicz in [29] defines the algorithm for the single arithmetical crossover in the following way:

Let the selected parents in the generation t be: $S_v^t = \langle V_1,...,V_m \rangle$ and $S_w^t = \langle W_1,...,W_m \rangle$. If the crossing over happens after the kth component then the resulting offspring will have the following form: $S_v^{t+1} = \langle V_1,...,V_k',...,V_m \rangle$ and $S_w^{t+1} = \langle W_1,...,W_k',...,V_m \rangle$ where $V_k' = a \cdot W_k + (1-a) \cdot V_k$ and $W_k' = a \cdot V_k + (1-a) \cdot W_k$. The constant a is defined as:

$$a \in \begin{cases} [\max(\alpha,\beta), \min(\gamma,\delta)], & \text{if } V_k > W_k, \\ [0,0], & \text{if } V_k = W_k, \\ [\max(\gamma,\delta), \min(\alpha,\beta)], & \text{if } V_k < W_k, \end{cases}$$

$$\alpha = \frac{l_{(k)}^{S_w} - W_k}{V_k - W_k}, \qquad \beta = \frac{u_{(k)}^{S_v} - V_k}{W_k - V_k}, \qquad (3.23)$$

$$\gamma = \frac{l_{(k)}^{S_v} - V_k}{W_k - V_k}, \qquad \delta = \frac{u_{(k)}^{S_w} - W_k}{V_k - W_k}.$$

where l and u are the lower and the upper bound for the kth component of the s_v and s_w individuals. It is important to note that the coefficient a value is determined separately for each single arithmetical crossover.

3.5.4.4 Whole arithmetical crossover

Michalewicz in [29] defines the whole arithmetical crossover as a linear combination of two parents in the following way:

Let the selected parents in the generation t be: $S_v^t = \langle V_1, \ldots, V_m \rangle$ and $S_w^t = \langle W_1, \ldots, W_m \rangle$. The resulting offspring are generated using the following linear combination: $s_v^{t+1} = a \cdot s_w^t + (1-a) \cdot s_v^t$ and $s_w^{t+1} = a \cdot s_v^t + (1-a) \cdot s_w^t$. The coefficient $a \in [0,1]$ is a simple static system parameter which guarantees that the offspring will stay within the convex solution space S.

3.5.4.5   Heuristic crossover

Michalewicz in [29] defines the heuristic crossover. This is a unique crossover operator for the following reasons:

− it uses the merit function values in determining the direction of the search;
− it produces only one offspring;
− it may produce no offspring at all.

The heuristic crossover is defined in the following way:

Let the selected parents in the generation t be: $S_v^t = \langle V_1, \ldots, V_m \rangle$ and $S_w^t = \langle W_1, \ldots, W_m \rangle$. The resulting single offspring is generated using the following equation:

$$S^{t+1} = r \cdot \left(S_w^t - S_v^t \right) + S_w^t \tag{3.24}$$

where r is a random number between 0 and 1, and the parent S_w^t is not worse than S_v^t, i.e. $\psi(s_w) \geq \psi(s_v)$ for the maximization problems and $\psi(s_w) \leq \psi(s_v)$ for the minimization problems. It is possible for this crossover operator to generate offspring, which is not feasible. In such a case another random number r is generated and another offspring created. If after ω attempts no new offspring meeting the constraints is found, the crossover operator gives up and produces no offspring.

3.5.5   Mutation

Mutation is traditionally in genetic algorithms community considered as the less important genetic operator which does not contribute in search for optimal solution. The primary role of the mutation is to keep diversity in the population. The mutation ought to create small changes in the individuals from the population. This point of view differs from the traditional positions of other evolutionary computation methods, such as evolution strategies and evolutionary programming. There mutation plays the main role in the search of optimization space and the optimum solution finding. It ought to be noticed that different evolutionary algorithms have heterogeneous definitions of the mutation as the random change of the individual's single component. The representation of the individual defines the type of random change. If the individual is represented with the bit strings then the random change is very simple – the single bit changes its value from 1 to 0 or vice versa. If the individual is represented with the floating point numbers then the mutation randomly chooses a numerical value from the given interval.

It is possible to define several mutation operators if the floating point numbers are used for the representation of the individuals from the population. Davis in [34] and Michalewicz in [29] described the following mutation operators:

– the floating point number mutation or the uniform mutation;
– the floating point number creep;
– the boundary mutation;
– the non-uniform mutation.

3.5.5.1 Floating point number mutation

Davis in [34] and Michalewicz in [29] described the same mutation operator, but they gave it different names. Davis called it the floating point number mutation and Michalewicz called it the uniform mutation. The definition of the mutation operator is following:

Let the selected parent for the mutation in the generation t be: $S_v^t = \langle V_1,...,V_m \rangle$. If the mutation takes place at the kth component then the resulting offspring will have the following form: $S_v^{t+1} = \langle V_1,...,V_k',...,V_m \rangle$ where V_k' is a random value with the uniform probability distribution from the range $[l,u]$. The lower bound l and the upper bound u of the range is defined by the optimization problem constraints.

3.5.5.2 Floating point number creep

Davis in [34] defined the floating point number creep as the mutation operator that moves the component of the individual for a small random value if the probability test is passed. The idea behind the floating point number creep is following: the individual that is reproducing is probably a very good individual and needs only to be improved. The optimization process can be visually described as a sequence of hills and valleys. If the optimization procedure is on a good hill, then it is desired to stay at that hill and keep descending to the valley in the case of minimization.

The mathematical definition of the floating point number creep is:

Let the selected parent for the mutation in the generation t be: $S_v^t = \langle V_1,...,V_m \rangle$. If the mutation takes place at the kth component then the resulting offspring will have the following form: $S_v^{t+1} = \langle V_1,...,V_k',...,V_m \rangle$ where $V_k' = a \cdot V_k$ is the kth component multiplied by the small random number.

3.5.5.3 Non-uniform mutation

Michalewicz in [29] defined the non-uniform mutation as the mutation operator responsible for the fine-tuning capabilities of the system. The mathematical definition of the non-uniform mutation is:

Let the selected parent for the mutation in the generation t be: $S_v^t = \langle V_1, ..., V_m \rangle$. If the mutation takes place at the kth component then the resulting offspring will have the following form: $S_v^{t+1} = \langle V_1, ..., V_k', ..., V_m \rangle$ where:

$$V_k' = \begin{cases} V_k + \Delta\left(t, u_{(k)}^{S_v^t} - V_k\right) & \text{if the random digit is } 0 \\ V_k - \Delta\left(t, V_k - l_{(k)}^{S_v^t}\right) & \text{if the random digit is } 1 \end{cases} \qquad (3.25)$$

where:

$u_k^{S_v^t}$ is the maximal allowed value of the kth component for the individual S_v^t in the generation t;

$l_k^{S_v^t}$ is the minimal allowed value of the kth component for the individual S_v^t in the generation t;

The function $\Delta(t, y)$ returns a value in the range $[0, y]$ such that the probability of $\Delta(t, y)$ being close to zero increases as the generation t increases. This property causes this operator to initially (when the t is small) search the space uniformly and very locally at later stages. Michalewicz uses the following function for $\Delta(t, y)$:

$$\Delta(t, y) = y \cdot \left(1 - r^{\left(1 - \frac{1}{T}\right)^b}\right) \qquad (3.26)$$

where:

r is a random number from $[0,1]$;

T is the maximal generation number;

b is a system parameter determining the degree of nonuniformity.

3.6 Parameters for genetic algorithms

The forth decision to make in implementing the GA optimization is how to set the values for the various parameters, such as the population size, the crossover rate, and the mutation rate. These parameters typically interact with one another nonlinearly, so they cannot be optimized one at a time. There is a great number of research about the parameter settings in genetic algorithms and their adaptation. No conclusive results on what is the best can be formulated. Most of the researchers use what has worked well in previously reported cases. In this section some experimental results in search for the "best" parameter settings will be reviewed.

De Jong in [27] performed an early systematic study of how varying parameters affected the performances of the GA optimization. He used the classical genetic algorithm with bit string representation of individuals in the population. The performance of the GA optimization was tested on a small set of specially designed functions. De Jong's experiments indicated that the best population size was 50-100 individuals, the best single point crossover rate was ~0.6 per pair of parents, and the best mutation rate was 0.001 per bit. These settings become widely used in the GA community, even though it was not clear how well the GA optimization would perform with these settings on the problems outside the De Jong's test suite.

Grefenstette in [42] noted that the genetic algorithm could be used to optimize parameters for another genetic algorithm. In Grefenstette's experiments, the "meta-level GA" evolved a population of 50 GA parameter sets for the problems in the De Jong's test suite. Each individual encoded six GA parameters: the population size, the crossover rate, the mutation rate, the generation gap, the scaling window and the selection strategy (elitist or no elitist). The results of the Grefenstette's experiments showed that the optimum population size is 30 individuals, the single point crossover rate (probability) is 0.95 and the bit mutation rate is 0.01. It must be noticed that these parameter values are valid for only functions defined in the De Jong's test suite and there are other functions for which these parameter settings are not optimal.

Shaffer *et al.* in [43] spent over a year of CPU time systematically testing a wide range of parameter combinations for the genetic algorithms. They used the modified genetic algorithm where the individuals were represented by the Gray codes (special way of representing bit strings). The set of test functions consisted of the small set of numerical optimization problems including some of the De Jong's functions. The obtained results are following: the population size 20 – 30 individuals, the crossover rate 0.75 – 0.95 and the mutation rate 0.005 – 0.01.

It seems unlikely that any general principles about parameter settings can be formulated *a priori*, in view of variety of problem types, encodings and performance criteria that are possible in different applications. Moreover the optimum population size, the crossover rate and the mutation rate are likely to change over the course of a single run. Many researchers consider that the most promising approach is to have the parameter values *adapt* in real time to the ongoing optimization. Davis in [34] described the genetic operator self-adaptation. He called his version of genetic algorithms ASSGA - Adaptive Steady-state Genetic Algorithm and it is described in the next section.

3.7   Adaptive steady-state genetic algorithm

Davis in [34] described the adaptive steady-state genetic algorithm, which is used for the construction of the genetic algorithm for the optical systems optimization. He defined the exact methodology for the genetic algorithm definition. Each genetic algorithm consists of three modules:

− the evaluation module;
− the population module;
− the reproduction module.

The evaluation module is the most important module because it defines the individual's representation, the boundary conditions which each individual must fulfill, the way of calculation and the form of the merit function. This module must be formed separately for each optimization problem.

The population module takes care about all parameters which are important for the definition of GA.

- the representation technique (binary or floating point);
- the initialization technique (usually random number initialization);
- the delatation technique (whether are deleted all individuals or the worst individual is deleted);
- the reproduction technique (generational – one whole generation is replaced by other whole generation or steady-state where only the worst individual is replaced);
- the parent selection technique (usually the roulette wheel parent selec-tion; all methods for the parent selection are described in section 3.4);
- the methods for the merit function scaling (all methods for the parent selection are described in section 3.3);

The reproduction module takes care about parameters which need to be defined for various types of genetic operators:

- the genetic operator selection technique (especially important ifthere are several different genetic operators);
- the genetic operator definition.

Davis in [34] defined the adaptive steady-state genetic algorithm - ASSGA as a genetic algorithm for the technical systems optimization. His genetic algorithm definition differs from the majority of genetic algorithms which are based on the research of Holland, De Jong and Goldberg.

The ASSGA will be defined according to the methodology defined in [34]. The evaluation module will not be described in this section because it is specific for each optimization problem.

The population module is:

- the representation technique: the floating point number representation;
- the initialization technique: the random number initialization;
- the delatation technique: delete only the individual with the worst merit function;
- the reproduction technique: the steady-state without duplicates;
- the parent selection: the roulette wheel parent selection;
- genetic operators: the uniform crossover, the average crossover, the floating point number mutation, the big floating point number creep, the small floating point number creep.

After discussion on various parameterization techniques Davis in [34] concludes that the best technique is to allow the genetic operators to change their weighting factors during the optimization. The idea behind this adaptive genetic operator weighting factors is following: the weighting factor size is directly proportional to the performance of any genetic operator over a recent interval. The performance of the genetic operator is measured by the number of very good individuals that are produced. If the genetic operator has produced or contributed to the production of a very good individual over a recent interval then it is assigned greater weighting factor. If the genetic operator has not produced or contributed to the production of a very good individual over a recent interval then it is assigned a smaller weighting factor. The greater weighting factor means that the genetic operator is selected more often to produce new offspring. The precise algorithm for the adaptive genetic operator weighting factors has three parts. The first part is a technique for measuring the genetic operator performance. Put intuitively, the genetic operator's performance is the amount of improvement in the population the genetic operator is responsible for, divided by the number of offspring the genetic operator created. The algorithm for computing the genetic operator performance is:

Step 1: Whenever an offspring is created, record who its parents were and the genetic operator that creates it.

Step 2: Whenever an offspring better than the best individual in the population is created, give it an amount of credit equal to the amount that its merit function exceeds the best individual's merit function.

Step 3: When the offspring is given credit for having a better merit function, add the portion of that credit to the credit amounts of the offspring parents, then to their parents, and so on. The portion of that credit to be passed back and the number of generations to pass back are parameters of this algorithm.

Step 4: To compute the performance of a genetic operator over an interval of a certain length, sum the credit of each offspring the operator produced in that interval and divide by the number of offspring. The length of the interval is a parameter of this algorithm.

The second part is a technique for the adapting genetic operator weighting factors. The algorithm consists of the following steps:

Assume that the genetic operator weighting factor is always 100. Let x be the amount of the genetic operator weighting factor to be adapted. Then:

Step 1: If all genetic operator performance measures are 0, leave the genetic operator weighting factors as they are.

Step 2: Otherwise multiply the list of genetic operator weighting factors by a constant factor so their sum is (100 - x). Call this new list of weighting factors the base weighting factors.

Step 3: Form a list of genetic operator performance measures over the recent interval.

Step 4: Multiply the list of genetic operator performance measures by a constant factor so that their sum is x. Call this list the list of adaptive weighting factors.

Step 5: Sum each genetic operator's base weighting factor and its adaptive weighting factor to derive a new weighting factor.

One potential problem with this algorithm is that a genetic operator that is important only at the end of an optimization may be lost because its weighting factor will become zero at the start of the optimization. Because of that the adaptation mechanism replaces the old value of the weighting factor for each genetic operator with its new value or minimum value, whichever is larger.

The third part is a technique for the initialisation of genetic operator weighting factors. The algorithm consists of the following steps:

Step 1: Set all genetic operator weighting factor values to be identical, unless the better initial values are known.

Step 2: Initialise the population and store away each individual and its merit function.

Step 3: Run the genetic algorithm on the initial population and these genetic operator weighting factor values to create x new individuals, where x is the size of the interval between adaptations of the weighting factor values.

Step 4: Adapt the genetic operator weighting factor values and store the adapted values away.

Step 5: Repeat steps 3-4 until the genetic algorithm is finished.

Step 6: Average the sets of adapted genetic operator weighting factor values derived in steps 1-5 and set the genetic operator weighting factor values to equal this average.

Step 7: Using these new genetic operator weighting factor values repeat steps 2-6.

Step 8: Continue replacing the genetic operator weighting factor value with the new value until the difference between the old values of genetic operator weighting factor and the new ones becomes negligible.

Chapter 4

Evolution Strategies

Evolution strategies (ES) are algorithms which imitate the principles of natural selection as a method to solve parameter optimization problems. Bienert, Rechenberg and Schwefel developed them in Germany during the 1960s. The evolution strategies can be divided in two large groups:

− the two membered evolution strategy;
− the multimembered evolution strategy.

4.1   Two membered evolution strategy

The two membered evolution strategy represents the minimal concept for an imitation of organic evolution. In the evolution theory three principles are considered as the most important ones: selection, recombination and mutation. The two membered evolution strategy uses only two principles mutation as sole genetic operator for the creation of new individuals and selection for the search of the optimization space and finding the best individual which represents the solution to the problem. It does not use the recombination because on the contrary to genetic algorithms in the evolution strategies it is considered that mutation is the main genetic operator. The individuals are represented with the floating point numbers.

Schwefel in [21] described the two membered evolution strategy using well known biological terms:

Step 1: Initialization

A given population consists of two individuals, one parent and one offspring. Only the parent needs to be specified as a starting point for the optimization.

Step 2: Mutation

The parent $E^{(g)}$ of the generation g produces an offspring $N^{(g)}$, which is slightly different from the parent.

Step 3: Selection

The parent and the offspring have a different capacity for survival in the same environment because they are different.

This is the simplest possible evolution simulation and the following assumptions are made:

- the population size remains constant;
- an individual has in principle an infinitely long life and capacity for producing offspring;
- only point mutations occur independently of each other at all single parameter locations;
- the environment and thus the criterion of survival is constant over time.

The same algorithm expressed in the mathematical terms is:

Step 1: Initialization

The optimization starts with the two points in the n-dimensional Euclidean space. Each point is characterized by a position vector consisting of a set of n components which are the floating point numbers.

Step 2: Mutation

Starting from the point $E^{(g)}$, with the position vector $x_E^{(g)}$ in the iteration g a second point $N^{(g)}$, with the position vector $x_N^{(g)}$ is generated using the equation $x_N^{(g)} = x_E^{(g)} + z^{(g)}$ with components $x_{N,i}^{(g)} = x_{E,i}^{(g)} + z_i^{(g)}$ where $z_i^{(g)}$ are random numbers, mutually independent.

Step 3: Selection

Two points described in the previous step have different merit function values $F(x)$. The point which has better (for the minimization smaller) merit function value can serve as a starting point for the new mutation in the next generation $g + 1$.

Go to step 2 as long as the termination criterion does not hold.

The main problem in this algorithm is how to choose the random vectors $z^{(g)}$. This choice has the role of mutation. Schwefel in [21] has decided to consider the mutations as the sum of the numerous individual events where the small changes occur frequently, but large ones only rarely. This is in accordance with the central limit theorem of statistics. Schwefel has decided to use the Gaussian normal distribution with the following basic parameters:

- the mean value ξ_i for a component z_i has the value zero;
- the standard deviation σ_i is small.

The probability density function for the normally distributed random events is:

$$\omega(z) = \frac{1}{\sqrt{2 \cdot \pi} \cdot \sigma_i} \cdot \exp\left(-\frac{(z_i - \xi_i)^2}{2 \cdot \sigma_i^2}\right) \tag{4.1}$$

If $\xi_i = 0$ is adopted then the well known $(0, \sigma_i^2)$ normal distribution is obtained. There is still however a total of n free parameters $\{\sigma_i, i = 1(1)n\}$ with which to specify the standard deviations of the individual random components. By analogy with other, deterministic optimization methods, the standard deviation σ_i can be called optimization step lengths, in the sense that they represent average values of the lengths of the random optimization steps.

For the particular random vector $z_i = \{z_i, \quad i = 1\,(1)\,n\}$ with the independent $(0, \sigma_i^2)$ distributed components z_i, the probability density function is:

$$\omega(z_1, z_2, ..., z_n) = \prod_{i=1}^{n} \omega(z_i) = \frac{1}{(2 \cdot \pi)^{\frac{n}{2}} \cdot \prod_{i=1}^{n} \sigma_i} \cdot \exp\left(-\frac{1}{2} \cdot \sum_{i=1}^{n} \left(\frac{z_i}{\sigma_i}\right)^2\right) \tag{4.2}$$

or more compactly if $\sigma_i = \sigma$ for all $i = 1(1)n$,

$$\omega(z) = \left(\frac{1}{\sqrt{2 \cdot \pi \cdot \sigma}}\right)^n \cdot \exp\left(\frac{-z \cdot z^T}{2 \cdot \sigma^2}\right) \tag{4.3}$$

Generation of the normally distributed numbers on the computer is the process that involves the generation of the uniformly distributed random numbers and their transformation to the normally distributed random numbers. There are lots of very good computer algorithms for the generation of the uniformly distributed random numbers that are easy to implement and have large period of repetition. Schwefel in [21] uses the transformation rules of Box and Muller to generate two independent normally distributed random numbers with the average value zero and the standard deviation unity from two independent uniformly distributed random numbers in the range [0,1]:

$$\begin{aligned} z_1' &= \sqrt{-2 \cdot \ln Y_1} \cdot \sin(2 \cdot \pi \cdot Y_2) \\ z_2' &= \sqrt{-2 \cdot \ln Y_1} \cdot \cos(2 \cdot \pi \cdot Y_2) \end{aligned} \tag{4.4}$$

where:

Y_i is the uniformly distributed random numbers in the range [0,1];

z_i' is the normally distributed random numbers in the range [0,1].

To obtain a distribution with the standard deviation different from the unity the z_i' ought to be multiplied by the standard deviation σ_i.

$$z_i = \sigma_i \cdot z_i' \tag{4.5}$$

4.1.1   Optimization step length control

In the mathematical optimization problems, the variables can have various values that can differ very much. Thus the optimization step length must be continuously modified if the optimization algorithm is to be efficient. If the optimization step length is too low then the unnecessarily large number of iterations is executed. If the optimization step length is too high then the optimum can only be crudely approached and the optimization can get stuck in the local minimum far from the global minimum. Thus in all optimization methods the step length control is the most important part of the algorithm after the recursion formula and it is closely related to the convergence behaviour of the optimization algorithm.

Schwefel in [21] defines the following rule for controlling the size of the optimization step length:

The 1/5 success rule

From time to time during the optimization calculate the frequency if successes i.e. the ratio of the number of successes (when the individual with the better merit function is generated) to the total number of mutation. If the ratio is greater than 1/5 increase the standard deviation, otherwise decrease the standard deviation.

In many optimization problems this rule shows to be very effective in maintaining the highest possible speed of the convergence towards the optimum.

Schwefel in [21] gives a detailed mathematical analysis of the 1/5 success rule and as the result he states a more precise, from the mathematical point of view, formulation of the 1/5 success rule for the numerical optimization:

The 1/5 success rule

After every n mutations check how many successes (the merit function of offspring is better than the merit function of parent) have occurred over the preceding $10 \cdot n$ mutations. If this number is less than $2 \cdot n$, multiply the step lengths by the factor 0.85, otherwise divide the step lengths by 0.85.

The numerical optimization is usually executed on the digital computers with the finite number of significant digits. If the step lengths are constantly multiplied by the factor 0.85 they become smaller and smaller and eventually become zero after a finite number of multiplications. Every subsequent division by the factor 0.85 leaves the step length as zero. If this happens to one of the standard deviations σ_i, the affected variable x_i remains constant thereafter. The optimization continues only in a subspace of $\mathbf{R}^n$. To guard against that the step lengths evetually become zero it must fulfil the following conditions:

$$\sigma_i^{(g)} \geq \varepsilon_a \qquad \qquad \text{for all} \quad i = 1 \,(1)\, n$$

$$\sigma_i^{(g)} \geq \varepsilon_b \cdot \left| x_i^{(g)} \right| \qquad \text{for all} \quad i = 1 \,(1)\, n$$

(4.6)

where:

$\varepsilon_a > 0$ is the absolute value for the lower bound of the step length. It must be chosen large enough to be treated as different from zero within the accuracy of the computer used.

$1 + \varepsilon_b > 1$ is the lower bound to step sizes relative to the values of variables. It must be chosen large enough to be treated as different from 1.0 within the accuracy of the computer used.

4.1.2 Convergence criterion

In numerical optimization, if the calculations are made by computer, the rule for the optimization termination ought to be built in. Towards the minimum the step lengths and distances covered normally become smaller and smaller. A frequently used convergence criterion consists of ending the optimization when the changes in the constructional variables become zero or smaller than the predefined small value, in which case no further improvement in the merit function is made, or when the optimization step lengths have become zero or smaller than the predefined small value. This procedure has however one disadvantage that can be serious. Low values for the optimization step lengths occur not only when the minimum is nearby, but also if the optimization is moving through the imagined narrow valley. The optimization may then be practically stopped long before the minimum is found. The optimization step length is not the measure of the closeness to the optimum, rather it conveys the information about the complexity of the minimum problem: the number of the constructional variables and the narrowness of the imagined valleys encountered. The requirement $\sigma > \varepsilon$ or $\left\| x^{(g)} - x^{(g-1)} \right\| > \varepsilon$ for the continuation of the optimization is thus no guarantee for the sufficient convergence.

Alternatively the convergence criterion can be the change in the merit function value. If the difference in the merit function values in two successive iterations becomes zero (or smaller than the small predefined value) the optimization is terminated. But this condition can also be fulfilled far from the minimum if the imagined valley in which the minimum (i.e. the deepest point) is sought happens to be very flat in shape. Thus the best strategy is to use both convergence criteria (change of the optimization step length and the change of the merit function). If for example the change in merit function values is zero then the optimization step length control, which is different from zero, enables continuation of the optimization and the passage through the flat part of the valley.

The convergence criterion can be defined as:

– the optimization step length must be smaller than the sufficiently small value defined by Eq. (4.6);

– the merit function difference must be smaller than the sufficiently small value. It is defined in the absolute and relative form:

$$\psi\left(x_E^{(g-\Delta g)}\right) - \psi\left(x_E^{(g)}\right) \le \varepsilon_c$$

$$\frac{\left|\psi\left(x_E^{(g-\Delta g)}\right) - \psi\left(x_E^{(g)}\right)\right|}{\psi\left(x_E^{(g)}\right)} \le \varepsilon_d \qquad\qquad (4.7)$$

where:

$$\Delta g \ge 20 \cdot n$$
$$\varepsilon_c > 0$$
$$1 + \varepsilon_d > 1$$

The condition $\Delta g \ge 20 \cdot n$, where n is the number of the variable constructional parameters, is designed to ensure that in an extreme case the standard deviations are reduced or increased within the test period by at least the factor $(0.85)^{\pm 20} \approx 25$ in accordance with the 1/5 success rule. It is well known that the more variable constructional parameters are needed in the problem definition, the more slowly the optimization progresses. Thus the recommended procedure is to make a test every $20 \cdot n$ mutations.

4.2 Multimembered evolution strategies

The two membered evolution strategy is successful in application to many optimization problems. But there are certain types of problems that can not be successfully solved, because the 1/5 success rule permanently reduces the optimization step lengths without improving the rate of convergence to the optimum solution. This phenomenon of reducing the optimization step length without improving the rate of convergence occurs frequently if the constraints become active during the optimization. The possible remedy is to allow the standard deviations σ_i to be individually adjustable. This is done in the multimembered evolution strategies.

The population of only two members: the parent and the offspring represent the foundation for the evolution simulation. In order to reach a higher level of imitation of the evolutionary process the number of individuals must be increased.

Schwefel in [21] described the multimembered evolution strategies using well-known biological terms:

Step 1: Initialization

A given population consists of μ individuals (parents), which are generated randomly according to the Gaussian normal distribution.

Step 2: Mutation

Each individual parent produces $\dfrac{\lambda}{\mu}$ offspring on average so that a total of λ new offspring are available. There are two possible ways of the offspring creation: the mutation and the recombination. All descendants differ slightly from the parents.

Step 3: Selection

Only the μ best of the λ offspring become parents for the next generation.

The same algorithm expressed in the mathematical terms for the minimization problem is:

Step 1: Initialization

Define $x_k^{(0)} = x_{E_k}^{(0)} = \left(x_{k,1}^{(0)}, ..., x_{k,n}^{(0)} \right)^T$ for all $k = 1\,(1)\,\mu$. $x_k^{(0)} = x_{E_k}^{(0)}$ is the vector of the kth parent E_k, such that all the boundary conditions fulfilled $G_j\left(x_k^{(g+1)} \right) \geq 0$ for all $k = 1\,(1)\,\mu$ and $j = 1\,(1)\,m$. Set the generation counter to zero $g = 0$.

Step 2: Mutation

Generate $x_l^{(g+1)} = x_k^{(g+1)} + z^{(g \cdot \lambda + l)}$, such that all the boundary conditions fulfilled $G_j\left(x_l^{(g+1)} \right) \geq 0$, $j = 1\,(1)\,m$, $l = 1\,(1)\,\lambda$ where $k \in [1, \mu]$. $x^{(g+1)} = x_{N_l}^{(g+1)} = \left(x_{l,1}^{(0)}, ..., x_{l,n}^{(0)} \right)^T$ is the vector of lth offspring N_l, and $z^{(g \cdot \lambda + l)}$ is a normally distributed random vector with n components.

Step 3: Selection

Sort the $x_l^{(g+1)}$ for all $l = 1\,(1)\,\lambda$ so that the merit function $\psi\left(x_{l_1}^{(g+1)} \right) \leq \psi\left(x_{l_2}^{(g+1)} \right)$, for all $l_1 = 1\,(1)\,\lambda$ and $l_2 = 1\,(1)\,\lambda$. Assign $x_k^{(g+2)} = x_{l_1}^{(g+1)}$, for all $k, l_1 = 1\,(1)\,\mu$. Increase the generation counter $g \leftarrow g + 1$. Go to step 2 unless some termination criterion is fulfilled.

In the algorithm the following variables are used:

μ is the number of parents;

λ is the number of offspring;

m is the number of boundary conditions.

4.2.1 Optimization step length control

The fundamental question that is posed is how to proceed in order to achieve the maximum rate of convergence to the optimum solution and to maintain the optimum values for the optimization step lengths in the case of the multimembered evolution strategies. For the two membered ES this aim was met by the 1/5 success rule. This is an outside control parameter and it does not correspond to the biological paradigm. In the case of the multimembered ES, the natural process is tried to be simulated as much as possible so the optimization step length control is done inside the algorithm i.e. in the same way as the optimization is done.

Each parent besides the variable constructional parameters $x_{E,i}$, $i = 1\,(1)\,n$ has a set of parameters $\sigma_{E,i}$, $i = 1\,(1)\,n$, which describes the optimization step lengths i.e. the standard deviations of the random changes. Each offspring N_l of the parent E should differ from its parent both in the variable constructional parameters vector $x_{l,i}$ and in the standard deviation vector $\sigma_{l,i}$. The standard deviation changes should also be random and small. Whether an offspring will become a parent in the next generation depends on its merit function and therefore only from the variable constructional parameters vector $x_{l,i}$. The values of the variable constructional parameters depend not only on the variable constructional parameters from the parent x_{Ei} but also on the standard deviations $\sigma_{l,i}$, which affect the size of changes of the variable constructional parameters $z_i = x_{l,i} - x_{E,i}$. In this way the standard deviations or in other words the optimization step lengths play an indirect role in the selection mechanism.

Schwefel in [21] concludes that the highest possible probability that an offspring is a better then the parent is:

$$\omega_{l\,max} = 0.5 \tag{4.8}$$

In order to prevent that the reduction of the standard deviations σ_i always gives rise to the selection advantage, the number of offspring must be at least $\lambda \geq 2$. But the optimal optimization step lengths can only take effect if:

$$\lambda > \frac{1}{\omega_{opt}} \tag{4.9}$$

This means that on average at least one offspring represents an improvement of the merit function value. The number of offspring per parent thus plays a decisive role in the multimembered ES, just as does check on the success ratio in the two memebered ES.

Schwefel in [21] recommends the following way of generating new optimization step lengths:

$$\sigma_N^{(g)} = \sigma_E^{(g)} \cdot \overline{z}^{(g)} \tag{4.10}$$

The median $\overline{\xi}$ of the random distribution for the quantity $\overline{z}$ must be equal to one to satisfy the condition that there is no deterministic drift without selection. Furthermore, an increase of the optimization step length should occur with the same frequency as a decrease, more precisely, the probability of occurrence of a particular random value must be the same as that of its reciprocal. The third requirement is that small changes should occur more often than large ones. All three requirements are satisfied by the lognormal distribution. Random quantities obeying this distribution are obtained from $(0, \tau^2)$ normally distributed numbers Y by the process:

$$\bar{z} = e^Y \tag{4.11}$$

The probability distribution for $\bar{z}$ is then:

$$\omega(\bar{z}) = \frac{1}{\sqrt{2 \cdot \pi} \cdot \tau} \cdot \frac{1}{\bar{z}} \cdot \exp\left(-\frac{(\ln \bar{z})^2}{2 \cdot \tau^2}\right) \tag{4.12}$$

where the τ is the standard deviation of the normally distributed random numbers Y.

Instead of only one common strategy parameter σ, each individual can now have a complete set of n different $\sigma_i, i = 1\,(1)\,n$ for every alternation in the corresponding n variable constructional parameters $x_i, i = 1\,(1)\,n$. Schwefel in [21] recommends the following scheme:

$$\sigma_{N,i}^{(g)} = \sigma_{E,i}^{(g)} \cdot \bar{z}_i^{(g)} \cdot \bar{z}_0^{(g)} \tag{4.13}$$

Schwefel in [21] concludes that the maximal speed of convergence is obtained if the number of offspring is at least five times greater than the number of parents ($\lambda \geq 5 \cdot \mu$).

4.2.2   Convergence criterion

The convergence criterion is similar to the two membered ES, which means that the changes in the merit function values are considered. For the multimembered ES the values of the variable constructional parameters and the optimization step lengths for each of the μ parents and the μ best offspring from the λ possible offspring are stored in each moment. In general the best individuals will differ in the variable construction parameters values and therefore in the merit function values until the optimum is found. This analysis enables definition of the simple convergence criterion.

From the population of μ parents $E_k, k = 1\,(1)\,\mu$ let the F_b be the best merit function value, for the minimization problem minimum:

$$\psi_b = \min_k \left\{ \psi\!\left(x_k^{(g)}\right), \quad k = 1\,(1)\,\mu \right\}$$

and ψ_w the worst merit function value, for the minimization problem maximum:

$$\psi_w = \max_k \left\{ \psi\!\left(x_k^{(g)}\right), \quad k = 1\,(1)\,\mu \right\}$$

then for ending the optimization the criterion is:

$$\psi_w - \psi_b \leq \varepsilon_c$$

$$\frac{\mu}{\varepsilon_d} \cdot \left(\psi_w - \psi_b\right) \leq \left| \sum_{k=1}^{\mu} \psi\!\left(x_k^{(g)}\right) \right| \tag{4.14}$$

where ε_c and ε_d are to be defined such that:

$$\begin{aligned} \varepsilon_c &> 0 \\ 1 + \varepsilon_d &> 1 \end{aligned} \tag{4.15}$$

From Eq. (4.14) it is seen that either absolutely or relatively the merit function values of the parents in a generation must be close before it can be concluded that the optimization is converged to the solution.

4.2.3   Genetic operators

The multimembered ES opens the possibility for imitating the sexual reproduction, which is the additional principle of the organic evolution that is very important for the numerical optimization problems. By recombination of the two parents a new resource for the changes is added to the only existing genetic operator – the point mutation. The fact that only a few most primitive organisms do not use recombination mechanism testifies that the recombination is very important in the evolution process.

There are three variants of the multimembered ES called: GRUP, REKO and KORR. The GRUP strategy uses only the point mutation as the genetic operator. This is the first multimembered ES. The REKO strategy evolved from the GRUP strategy and uses two kinds of recombination and the point mutation as genetic operators. The KORR strategy is the most advances ES and uses several kinds of recombination which can be divided in the sexual, panmictic and random and the point mutation.

Michalewicz in [29] defines the point mutation as:

Let the randomly selected parent is defined as a pair of vectors $v = (\mathbf{x}, \sigma)$ where the $\mathbf{x}$ is the variable constructional parameters vector and the σ is the optimization step length vector i.e. the vector with the standard deviations. The mutation is defined with the following equations:

$$\sigma' = \sigma \cdot e^{N(0,\Delta\sigma)}$$
$$x' = x + N(0,\sigma')$$

(4.16)

where:

$\Delta\sigma$ is the input parameter of the method;

$N(0,\Delta\sigma), N(0,\Delta\sigma')$ is the vector of the independent random numbers according to the Gaussian normal distribution law.

Michalewicz in [29] defines the recombination operators used in the REKO strategy. In the REKO strategy it is possible to use two recombination operators: discrete and intermediate recombination.

The discrete recombination in ES is similar to the uniform crossover in the genetic algorithms. It is defined as:

Let the two parents are chosen randomly:

$$(x^1,\sigma^1) = [(x^1_1,...,x^1_n), (\sigma^1_1,...,\sigma^1_n)]$$
$$(x^2,\sigma^2) = [(x^2_1,...,x^2_n), (\sigma^2_1,...,\sigma^2_n)]$$

when the discrete recombination is applied the one offspring is obtained:

$$(x,\sigma) = [(x^{q_1}_1,...,x^{q_n}_n), (\sigma^{q_1}_1,...,\sigma^{q_n}_n)]$$

(4.17)

where indices q_i can have values 1 or 2 i.e. each component of the offspring is randomly chosen from the first or the second parent.

The intermediate recombination in ES is similar to the average crossover in the genetic algorithms. It is defined as:

Let the two parents are chosen randomly:

$$(x^1,\sigma^1) = [(x^1_1,...,x^1_n), (\sigma^1_1,...,\sigma^1_n)]$$
$$(x^2,\sigma^2) = [(x^2_1,...,x^2_n), (\sigma^2_1,...,\sigma^2_n)]$$

when the intermediate recombination is applied the one offspring is obtained:

$$(x,\sigma) = \left[\left(\frac{x^1_1+x^2_1}{2},...,\frac{x^1_n+x^2_n}{2}\right), \left(\frac{\sigma^1_1+\sigma^2_1}{2},...,\frac{\sigma^1_n+\sigma^2_n}{2}\right)\right]$$

(4.18)

as can be seen from Eq. (4.18) each component of the offspring represents the arithmetical mean value of the selected parents components.

In the KORR strategy several recombination genetic operators are possible. Bäck in [31] divides them in two categories: sexual and panmictic. The sexual recombination operators act on two individuals randomly chosen from the parent population. For the panmictic recombination operators one parent is randomly chosen and held fixed while the second parent is randomly chosen anew from the complete population for each component of the first parent. In other words, the creation of a single offspring individual may involve up to all parent individuals. This method for recombination emphasizes the point of view that the parent population as a whole forms a gene pool from which new individuals are created. The author in his research further modified the panmictic recombination operators. He allowed that each component of the both parents is randomly selected from the whole population. These recombination operators will be called random.

In the KORR strategy not only the variable constructional parameters are subject to the recombination but also strategy parameters (standard deviations and rotation angles). The recombination operator may be different for the variable constructional parameters, standard deviations and rotation angles. This implies that the recombination of these groups of information proceeds independently of each other.

In general there are two kinds of recombination: discrete and intermediate. In the KORR strategy there are the following variants:

- **the discrete recombination** for each component of the offspring: it is randomly chosen which of the two selected parents will contribute.

- **the discrete panmictic recombination** for each component of the offspring: one parent is randomly selected from the whole population and the other parent is fixed. The new component is randomly chosen from these parents.

- **the discrete random recombination** for each component of the offspring: the two parents are randomly selected from the whole population. The new component is randomly chosen from these parents.

- **the intermediate recombination** for each component of the offspring: the arithmetical mean value of the two selected parent components is calculated.

- **the intermediate panmictic recombination** for each component of the offspring: one parent is randomly selected from the whole population and the other parent is fixed. The new component is obtained as the arithmetical mean value of the components from these parents.

- **the intermediate random recombination** for each component of the offspring: the two parents are randomly selected from the whole population. The new component is obtained as the arithmetical mean value of the components from these parents.

- **the generalized intermediate recombination** In the generalized intermediate recombination the arbitrary weight factors from the interval [0, 1] can be chosen instead of the weight factor 0.5 (arithmetical mean).

$$x_i' = x_{1,i} + \chi \cdot \left(x_{2,i} - x_{1,i}\right) \tag{4.19}$$

where:

x_i' is the ith component of the offspring;

$x_{1,i}$ is the ith component of the first parent;

$x_{2,i}$ is the ith component of the second parent;

χ is the weight factor from the interval [0,1].

- **the panmictic generalized intermediate recombination** is the same with the generalized intermediate recombination except that the ith component of the second parent is chosen anew from the whole population for each component and the weight factor is chosen also anew for each component.

- **the random generalized intermediate recombination** uses the same Eq. (4.19) but the ith components of the first and the second parent are chosen anew from the whole population for each component and the weight factor is chosen also anew for each component.

4.2.4   Selection

The selection operators used in ES are completely deterministic in contrast to the selection operators in the genetic algorithms, which are heuristic. Schwefel in [21] introduced an elegant notation for the selection mechanisms in ES, characterizing the basic method and the number of parents and offspring respectively.

Because the population consists from the μ parents and the λ offspring it is possible to define two strategies for the selection of the μ parents for the next generation:

- **"plus strategy" - (μ+λ) ES**

   The μ parents for the next generation are chosen as the μ best individuals from the set of μ+λ individuals i.e. the set is constructed from all the parents and all the offspring.

- **"coma strategy" - (μ, λ) ES**

   The μ parents for the next generation are chosen as the μ best individuals from the λ offspring. The necessary condition is $\lambda > \mu$.

At the first glance the (μ+λ) selection with its guaranteed survival of the best individuals seems to be more effective. But Bäck in [31] gives the following reasons why the (μ, λ) selection is better than the (μ+λ) selection:

- In case of changing environments the (μ+λ) selection preserves (outdated) solutions and is not able to follow the moving optimum;

- The possibility of a (μ, λ) selection to forget good solutions in principle allows for leaving (small) local optima and is therefore advantageous in the case of multimodal topologies.

- The (μ+λ) selection hinders the self-adaptation mechanism with respect to strategy parameters to work effectively, because misadapted strategy parameters may survive for a relatively large number of generations when they cause a fitness improvement by chance.

Schwefel in [21] and Bäck in [31] show that the optimum ratio of the offspring and the parents is $\lambda \geq 6 \cdot \mu$ for example $\lambda = 7 \cdot \mu$ and the number of parents in the generation $\mu = 15$.

Summarizing the conditions for a successful self-adaptation of strategy parameters the following list is obtained:

- The (μ, λ) selection is required in order to facilitate extinction of maladapted individuals.
- The selective pressure may not become too strong i.e. the number of parents is required to be clearly larger than one.
- The recombination of strategy parameters is necessary (usually, intermediate recombination gives the best results).

Bäck in [31] shows that the lower values for the number of parents lead to the increasing in the convergence speed of the algorithm and the higher values for the number of parents lead to a more detailed search of the optimization space and to the decreasing in the convergence speed of the algorithm.

Chapter 5

Comparison of optimization algorithms

Large number of optimization algorithms necessarily raises the question about the best optimization algorithm. There seems to be no unique answer. It is logical that if an optimal optimization algorithm exists, then all other optimization algorithms would be superfluous. It is clear that the universal optimization algorithm, that can solve all problems occurring in practice, does not exist. All the optimization algorithms presently known can only be used without restriction in particular areas of application. According to the nature of a particular problem one or another optimization algorithm offers a more successful solution.

5.1 Comparison of classical and modern evolutionary optimization algorithms

The classical optimization algorithms, which are based on the damped least squares optimization, are described in details in Chapter 2. The modern optimization algorithms, based on the analogy with the natural evolution, are:
– the genetic algorithms described in details in Chapter 3;
– the evolution strategies described in details in Chapter 4.

The most fundamental difference between these optimization algorithms is that the classical optimization can find only the local optimum and the evolutionary algorithms try to find the global optimum. It cannot be proved that the evolutionary algorithms always find the global optimum but they find sufficiently good solutions close to the global optimum. What this means is that the classical optimization algorithms do not have possibility of escaping the first local optimum they find because they do not allow worsening of the merit function. To make it even more clear let us imagine the optimization as a try to find the highest peak in the mountain by a blind man. In order to find the highest peak he always ought to go uphill and never downhill. When he comes to the point that he no more can go uphill he concludes that this is the highest peak of the mountain. In the mathematics this is called the local optimum. The blind man does not know if there is a higher peak in the neighbourhood that is not reachable to him because he can only go uphill and never downhill.

The evolutionary algorithms, on the contrary to the classical optimization algorithms, have a possibility of escaping the local optimum and keep looking for the better local optimum and hopefully the global optimum. This is done by letting the temporary worsening of the merit function.

The second important difference is the selection of the starting point for optimization. For the classical optimization it is of utmost importance to select the appropriate starting point from which it is possible to go directly to the good local optimum. The selection of the starting point is directly connected with the optimum finding. To select a good starting point the optical designer needs experience, intuition, practice and lot of theoretical knowledge. The most experienced optical designer cannot be sure that he has selected the best starting point or the starting optical system for the optimization. Because of that it is usual to optimize several different starting optical systems just to see which one will give the best-optimized design.

With the evolutionary algorithms the connection between the starting point and the found optimum design is not so direct. The optical designer can select more easily the starting optical system and the evolutionary algorithms will randomly fill the rest of the optical system population. These optimization algorithms have a well known property of escaping the local optimum and searching all the optimization space for the global optimum.

The classical optimization methods give the best results when the good starting point or the optical system is known in advance. The classical optimization has the quick convergence to the nearest local optimum. The classical optimization is the order of magnitude faster in finding the optimum solution than the evolutionary algorithms.

The evolutionary algorithms are best suited when the good starting point is not known or when it is needed to search the whole optimization space. The evolutionary algorithms have slower convergence but they are able to search parts of the optimization space that usually cannot be reached by the classical optimization.

5.2   Comparison of the evolution strategies and genetic algorithms

The basic difference between the evolution strategies and the genetic algorithms lies in their domains. The evolution strategies were developed as methods for the numerical optimization. The genetic algorithms were formulated as a general-purpose adaptive search technique, which allocate exponentially increasing number of trials for above-average individuals. The genetic algorithms are applied in a variety of domains and a parameter optimization was just one field of their application.

The major similarity between the evolution strategies and the genetic algorithms is that both systems maintain population of potential solutions and make use of the selection principle of the survival of fitter individuals. However, there are many differences between these approaches.

The first difference between the evolution strategies and the classical genetic algorithm is in the way they represent the individuals. ES operates on the floating point vectors and the classical GA operates on the binary vectors. The author used the adaptive steady-state genetic algorithm (ASSGA) which is specially developed genetic algorithm for the numerical optimization. It also uses the floating point vectors.

The second difference between the genetic algorithms and the evolution strategies is in the way the selection is performed. In the single generation of the evolution strategies, μ parents generate intermediate population, consisting of λ offspring produced by means of the recombination and mutation operators for the (μ, λ) ES, plus the original μ parents for the $(\mu + \lambda)$ ES. Then the selection process reduces the size of this intermediate population back to μ individuals by removing the least fit individuals from the population. In a single generation of the genetic algorithm a selection procedure selects *pop_size* individuals from the *pop_size* – sized population. The individuals are selected with repetition i.e. a superior individual has a good chance to be selected several times to a new population. At the same time, even the weakest individual has a chance to be selected in the new population. The author used the ASSGA, which is different from the rest of genetic algorithms because it does not allow repetition. All individuals in the population must be different. Also in one generation only small number of individuals is selected for reproduction and mutation. All others are copied without modification to the next generation.

In the evolution strategies the selection mechanism is deterministic: it selects the best μ out of $\mu + \lambda$, for the $(\mu + \lambda)$ ES, or λ, for the (μ, λ) ES, individuals with no repetitions. In the classical genetic algorithms the selection procedure is random selecting *pop_size* out of *pop_size* individuals with repetition. The chance of selection for each individual is proportional to its fitness. The author used the ASSGA where the selection procedure is random selecting small number out of *pop_size* individuals without repetition for crossover and mutation. The rest of the population is copied without modification to the next generation.

The next difference between the evolution strategies and genetic algorithms is that reproduction parameters for the classical genetic algorithms (probability of crossover, probability of mutation) remain constant during the evolution process whereas the evolution strategies change them all the time. The author used the ASSGA where the reproduction parameters are updated after certain number of generations according to their performance.

ES and GA also handle constraints in a different way. Evolution strategies assume a set of $q > 0$ inequalities, $c_1(x) \geq 0, ..., c_q(x) \geq 0$, as a part of the optimization problem. If, during some iteration, an offspring does not satisfy all these constraints, then the offspring is disqualified i.e. it is not placed in a new population. The major strategy for the classical genetic algorithms for handling constraints is to impose penalties on individuals that violate them. The author used the ASSGA that has the following treatment of the constraints: if during the optimization an offspring does not satisfy all constraints then it is discarded and a new offspring is created and tested.

5.3   Numerical comparison of optimization algorithms

Schwefel in [21] conducted research in the numerical properties of the optimization algorithms both classical and modern. He worked with following algorithms:

– Coordinate strategy with Fibonacci search;
– Coordinate strategy with golden section;
– Coordinate strategy with Lagrangian interpolation;
– Direct search of Hooke and Jeeves;
– Davies – Swann – Campey method with Gram – Schmidt orthogonalization;
– Davies – Swann – Campey method with Palmer orthogonalization;
– Powell's method of conjugate directions;
– Stewart's modification of the Davidon – Fletcher – Powell method;
– Simplex method of Nelder and Mead;
– Method of Rosenbrock with Gram – Schmidt orthogonalization;
– Complex method of Box;
– (1+1) Evolution strategy EVOL;
– (10,100) Evolution strategy GRUP;
– (10,100) Evolution strategy with recombination REKO.

Schwefel in [21] performed following tests:

– the theoretical convergence rates for a quadratic objective function,
– the reliability of the optimization methods convergence on non-quadratic non-linear problems,
– the non-quadratic problems with many variables.

In this monography only the results of the evolution strategies will be described.

In the first group of numerical tests the theoretical and calculated convergence rates for a set of quadratic objective functions are presented. In all tests the number of variables are increased and the number of necessary iterations and the computational time is presented in Schwefel [21]. From the presented results it is clear that the evolution strategies have nearly linear dependence of the number of variables and the number of mutations or generations. The computational time needed to solve the problems is considerably higher than for classical optimization methods. It is noteworthy that the (10,100) evolution strategy without recombination only takes about 10 times as much as (1+1) evolution strategy, in spite of having to execute 100 mutations per generation. This factor of acceleration is significantly higher than the theory predictions.

The second group of tests represent a large group of non-quadratic non-linear problems. All this problems are divided in two large groups:

– the unconstrained problems,
– the constrained problems.

The very important part of all numerical optimization algorithms comparison is the test of its robustness. The robustness of the optimization method can be defined with the number of problems that are solved by a given algorithm. The results of all numerical optimization tests are presented in Schwefel [21]. For the group of unconstrained problems the evolution strategies are distinguished by the fact that in no case do they completely fail and they are able to solve far more than a half of all the problems exactly. The recombination in evolution strategies almost always improves the chance of getting very close to the desired solutions. Discontinuities in the partial derivatives and saddle points have no obvious adverse effects. On the whole the evolution strategies cope with constrained problems no worse than the classical optimization methods like Rosenbrock or complex methods, but they do reveal inadequacies that are not apparent in unconstrained problems. In particular the 1/5 success rule for adapting the variances of the mutation step lengths in the (1+1) evolution strategy appears to be unsuitable for attaining an optimal rate of convergence when several constraints become active. In problems with active constraints, the tendency of the evolution strategies to follow the average gradient trajectory causes the search to come quickly up against one or more boundaries of the feasible region. The subsequent migration towards the objective along such edges takes considerable effort and time.

5.4 Possibility of the classical optimization and the genetic algorithms hybridization

In many cases the effectiveness of a genetic algorithm can be enhanced by hybridizating with another optimization algorithm. One of the main benefits of the genetic algorithm paradigm is its domain independence i.e. the fact that it operates on a coding of a problem and needs to know nothing about the problem itself. Creating a hybrid optimization algorithm inevitably means that the domain independence characteristic is lost because some problem – specific knowledge is introduced, but this is often a price worth paying if the goal is to find an effective and efficient solution to a particular problem.

Hybridization can be carried out in various ways. For the optical system optimization it seems to be most appropriate to follow hybridization i.e. the combination of the genetic algorithm and the classical optimization. The author used the adaptive steady-state genetic algorithm (ASSGA) as the genetic algorithm for the optical system optimization and the damped least squares (DLS) for the classical optimization. The idea behind the hybridization is to use the ASSGA to find regions of solution space which are good candidates for locating the optimum and the classical optimization to actually find the optimum. This means that the ASSGA is used to find a set of different good starting points and the DLS starting from these points to find the optimum solution. This hybridization of the ASSGA and the DLS uses their good characteristics and hides their not so good characteristics.

One of the main drawbacks of the DLS is the importance of the good starting point selection i.e. selection of the good starting optical system. This usually means that the optical designer must use its experience and intuition to find good starting points. In designing the complex optical systems this means that less experienced optical designer will find a sub optimal optical system. The ASSGA is not so dependent on the starting optical system because it can escape from the local minimum and keep searching for the next better optical system. This means that the optical designer can chose more freely the starting optical system and that experience and intuition are not crucial for the starting optical system selection. This also means that a less experienced optical designer can design complex optical systems that were previously designed by the most experienced optical designers.

The genetic algorithms in general and the ASSGA in particular have very slow convergence rate to the optimum solution in comparison with the classical DLS optimization. So it makes sense to speed the convergence to the optimum solution by using the DLS optimization when the good starting optical systems are found by the ASSGA.

From the presented analysis it can be seen that the hybrid optimization algorithm will have two very desirable properties:

− the independence form the starting optical system,
− the speed of convergence to the optimal solution once the good starting optical systems are found.

Chapter 6

Ray trace

6.1 Introduction

Every optical system consists of one or more reflecting or refracting surfaces. The function of the optical system is to transform the diverging spherical wave fronts coming from object points in the object space to the converging spherical wave front going towards image points in the image space. The passage of the wave fronts through the optical system can be most easily described by using the concept of rays. The ray can be defined as a line normal to the wave and travelling in the direction of the advancing wave front. The passage of rays through the optical system may be determined by purely geometrical considerations, therefore it is correct to make following assumptions:

— a ray travels in a straight line in a homogeneous medium;

— a ray reflected at an interface obeys the low of reflection;

— a ray refracted at an interface obeys the law of refraction.

Computing the passage of rays through the optical system is a purely geometrical problem that can be solved by the analytic geometry methods.

Nearly every theoretical optical system consists of centred reflecting and refracting surfaces. In a centred optical system all surfaces are rotationally symmetrical about a single axis. A typical optical system, for example the Cooke triplet, consists from all spherical surfaces centres of which are assumed to lie on the optical axis. Production of these spherical surfaces and also the plane surfaces is a well-established technique and most commonly used. There are also aspherical surfaces which offer certain advantages in design but are much more difficult to produce. The aspherical surfaces also usually have rotational symmetry around the optical axis.

In ray tracing the plane surfaces can be considered to be special cases of spherical surfaces having the radii equal to infinity. For the aspherical surfaces ray tracing can be done by extending the technique for spherical surfaces. In the ray trace all surface types: the plane surfaces, the spherical surfaces and the aspherical surfaces are considered to be centred.

6.2   Types of rays and ray scheme

Before the description of the ray trace procedure is given, the various types of rays have to be defined. A general ray is any ray passing from any object point through the optical system to its image point on the image surface. The special ray that lies in the plane containing the optical axis and the object point (meridional plane) is called a meridional ray. There are several types of the meridional rays. The ray, which enters the optical system parallelly to the optical axis, is called an axial ray. The ray, which passes through the centre of the aperture diaphragm and forms the field angle with the optical axis, is called a principal ray. The ray, which is parallel to the principal ray and enters to the optical system at a certain height, is called an oblique ray. Any non-meridional ray is called a skew ray. The ray close to the optical axis and in the meridional plane is called a paraxial ray.

In order to understand clearly the kind of image formed by an optical system and what must be done to improve this image a certain number of rays must be determined in their passage through the optical system. These set of rays is usually called ray scheme and consists of several types of rays:

− the paraxial rays which are used to determine the basic optical system properties like the effective focal length, the back focal length and so on;

− the axial rays entering the optical system parallelly with the optical axis at several heights (usually: $0.5 \cdot h_{max}$, $0.7 \cdot h_{max}$ and h_{max});

− the principal rays entering the optical system with several field angles (usually: $0.5 \cdot \omega_{max}$, $0.7 \cdot \omega_{max}$ and ω_{max});

− the oblique rays entering the optical system with several field angles and at several heights (usually: $0.5 \cdot \omega_{max}$, $0.7 \cdot \omega_{max}$, ω_{max} and $0.5 \cdot h_{max}$, $0.7 \cdot h_{max}$, h_{max});

− the skew rays entering the optical system outside the meridional plane with several field angles and at several heights (usually: $0.5 \cdot \omega_{max}$, $0.7 \cdot \omega_{max}$, ω_{max} and $0.5 \cdot h_{max}$, $0.7 \cdot h_{max}$, h_{max}).

The axial, the principal, the oblique and the skew rays are used for the calculation of various aberration types. The aberrations are image errors and will be discussed in the next chapter.

6.3   Definitions and conventions

The ray tracing formulae to be used for tracing a ray through an optical system involves parameters of more than a single surface or a single medium. Therefore, it is important to adopt a notation convention, which will clearly distinguish one surface from another and one medium from another. Here are given following definitions and conventions which are in agreement with those given in MIL-STD-34:

1. It will be assumed that the light initially travels from left to right;

2. An optical system will be regarded as a series of surfaces starting with an object surface and ending with an image surface. The surfaces will be numbered consecutively, in the order in which the light is incident on them starting with zero for the object surface and ending with k for the image surface. A general surface will be called the jth surface.

3. All quantities between surfaces will be given the number of the immediately preceding surface.

4. A primed subscript will be used to denote quantities after refraction only when necessary.

5. r_j is the radius of the jth surface. It will be considered positive when the centre of curvature lies to the right of the surface.

6. The jth surface curvature is $c_j = \dfrac{1}{r_j}$. c_j has the same sign as r_j.

7. t_j is the axial thickness of the space between the jth and $(j+1)^{th}$ surface. It is positive if the physical ray is traveling from left to right. Otherwise it is negative.

8. n_j is the index of the refraction for the material between the jth and $(j+1)^{th}$ surface. It is positive if the physical ray is traveling from left to right. Otherwise it is negative.

9. K_j, L_j, M_j are the optical direction cosines. They are products of n_j and the direction cosines (with respect to the X, Y, Z axes respectively) of a ray in the space between the jth and $(j+1)^{th}$ surface.

10. The right-handed coordinate system will be used. The optical axis will coincide with the Z axis. The light travels initially toward larger values of the Z axis.

11. X_j, Y_j, Z_j are the position coordinates of a ray where it intersects the jth surface.

6.4 Ray trace equations for spherical surfaces

In the optical system ray travels in a straight line from a point on one surface to a point on the following surface. It is then refracted and proceeds to the next surface in a straight line. The basic ray trace procedure consists of the two parts: the transfer procedure and the refraction procedure. The transfer procedure involves computing the intersection point of the ray on the next surface from the optical direction cosines and the intersection point data at the previous surface. This is given $K_{j-1}, M_{j-1}, L_{j-1}$ and $X_{j-1}, Y_{j-1}, Z_{j-1}$ compute X_j, Y_j, Z_j. The equations used are called the transfer equations. The refraction procedure involves computing the optical direction cosines of a ray from the intersection point data and the optical cosines of the previous ray segment. That is, given X_j, Y_j, Z_j and $K_{j-1}, M_{j-1}, L_{j-1}$ compute K_j, L_j, M_j. The equations used are called the refraction equations.

After having applied the transfer and the refraction procedure, the initial data for the next surface is generated. The transfer equations will be used to compute

$X_{j+1}, Y_{j+1}, Z_{j+1}$ and the refraction equations will be used to compute $K_{j+1}, M_{j+1}, L_{j+1}$. It should be noted that it is often convenient to introduce fictitious or non-refracting surfaces to simplify the procedure. One example is the tangent plane, a XY plane tangent to a physical surface at the optical axis. These fictitious surfaces are handled in exactly the same manner as a physical surface. Transfer equations are used to go to or from such a surface.

6.4.1   Ray transfer procedure from the physical surface to the next tangent plane

The skew ray passage from one surface to another surface is shown in Fig. 6.1.

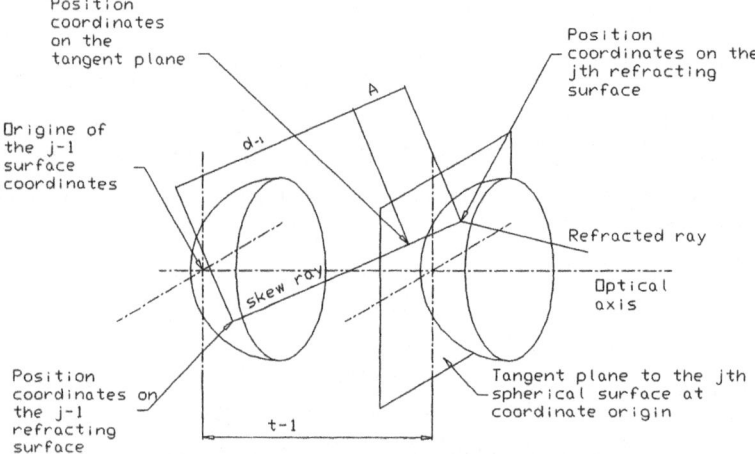

Figure 6.1. Diagram of a skew ray in the space between the $(j-1)^{th}$ surface and j^{th} surface

The initial data for the skew ray trace traversing the space between two surfaces is:

– the coordinates of the ray exit point on the $(j-1)^{th}$ surface: $X_{j-1}, Y_{j-1}, Z_{j-1}$;

– the optical direction cosines for the $(j-1)^{th}$ surface: $K_{j-1}, M_{j-1}, L_{j-1}$.

In order to facilitate the skew ray trace the tangent plane is positioned in the vertex of the spherical surface. Therefore the Z coordinate for all points of this plane is $Z_T = 0$. The coordinate X_T is obtained when the coordinate X_{j-1} is added to the coordinate ΔX which represents the skew ray projection, of length d_{j-1} onto the X axis.

$$X_T = X_{j-1} + \Delta X = X_{j-1} + d_{j-1} \cdot \frac{K_{j-1}}{n_{j-1}} \tag{6.1}$$

where the $\dfrac{K_{j-1}}{n_{j-1}}$ is the ray direction cosines with respect to the X axis.

There is a corresponding equation for the Y_T :

$$Y_T = Y_{j-1} + \Delta Y = Y_{j-1} + d_{j-1} \cdot \dfrac{L_{j-1}}{n_{j-1}} \tag{6.2}$$

The ray length d_{j-1} between the $j-1$ and the tangent plane is not given and must be calculated from the initial data. The change in the Z coordinate is given by:

$$\Delta Z = t_{j-1} - Z_{j-1} \tag{6.3}$$

and this equals to the ray projection along the Z axis:

$$\Delta Z = d_{j-1} \cdot \dfrac{M_{j-1}}{n_{j-1}} \tag{6.4}$$

Combining Eq. (6.3) with Eq. (6.4) the desired ray length is obtained as:

$$\dfrac{d_{j-1}}{n_{j-1}} = \left(t_{j-1} - z_{j-1} \right) \cdot \dfrac{1}{M_{j-1}} \tag{6.5}$$

6.4.2 Ray transfer procedure from the tangent plane to the next spherical surface

Since the tangent plane is not a refracting plane, the ray continues along the same direction until the spherical surface. Let the distance that the ray traverses is marked as A (as shown in Fig. 6.1) and there the ray has the same optical direction cosines as along the segment d_{j-1}. Therefore the new coordinate values X_j, Y_j, Z_j on the spherical surface can be determined from the coordinate values X_T, Y_T, Z_T on the tangent plane by Eq. (6.1) and Eq. (6.2). It is necessary to be taken in account that the $Z_T = 0$.

$$X_j = X_T + \dfrac{A}{n_{j-1}} \cdot K_{j-1}$$

$$X_j = X_T + \frac{A}{n_{j-1}} \cdot K_{j-1}$$

$$Y_j = Y_T + \frac{A}{n_{j-1}} \cdot L_{j-1} \tag{6.6}$$

$$Z_j = \frac{A}{n_{j-1}} \cdot M_{j-1}$$

In order to use Eq. (6.6) it is necessary to calculate the value of the distance A which is function of the jth spherical surface curvature, the ray coordinates at the tangent plane and the ray direction cosines. From the analytical geometry the equation of the sphere with the centre moved from the coordinate system origin for the value of radius r along the Z axis is given by:

$$c_j^2 \cdot \left(X_j^2 + Y_j^2 + Z_j^2\right) - 2 \cdot c_j \cdot Z_j = 0 \tag{6.7}$$

Substituting into Eq. (6.7) the expressions for X_j, Y_j, Z_j from Eq. (6.6) the following is obtained:

$$c_j \cdot \left(\frac{A}{n_{j-1}}\right)^2 \cdot \left(K_{j-1}^2 + L_{j-1}^2 + M_{j-1}^2\right) -$$

$$-2 \cdot \left(\frac{A}{n_{j-1}}\right) \cdot \left[M_{j-1} - c_j \cdot \left(X_T K_{j-1} + Y_T L_{j-1}\right)\right] + c_j \cdot \left(X_T^2 + Y_T^2\right) = 0 \tag{6.8}$$

If it is assumed that $c_j \neq 0$ i.e. that the spherical surface is not the plane surface then Eq. (6.8) can be divided by c_j. Since the sum of the optical direction cosines squares is:

$$K_{j-1}^2 + L_{j-1}^2 + M_{j-1}^2 = n_{j-1}^2 \tag{6.9}$$

and introducing following notations:

$$B = M_{j-1} - c_j \cdot \left(X_T \cdot K_{j-1} + Y_T \cdot L_{j-1}\right)$$

$$H = c_j \cdot \left(X_T^2 + Y_T^2\right) \tag{6.10}$$

Eq. (6.8) becomes:

$$c_j \cdot n_{j-1}^2 \cdot \left(\frac{A}{n_{j-1}}\right)^2 - 2 \cdot B \cdot \left(\frac{A}{n_{j-1}}\right) + H = 0 \tag{6.11}$$

The solution of the square equation (6.11) is:

$$\frac{A}{n_{j-1}} = \frac{B \pm n_{j-1} \cdot \sqrt{\left(\frac{B}{n_{j-1}}\right)^2 - c_j \cdot H}}{c_j \cdot n_{j-1}^2} \tag{6.12}$$

For the case of the plane surface ($c_j = 0$) the value of the distance A is zero because the tangent plane coincides with the plane surface.

The root $\sqrt{\left(\frac{B}{n_{j-1}}\right)^2 - c_j \cdot H}$ is very important for the calculation of the ray

transfer from the tangent plane to the spherical plane and for the refraction on the spherical plane. Therefore its physical meaning will be discussed. In order to be able to see its physical meaning it must be transformed into another form. To do this it is necessary to consider Fig. 6.2.

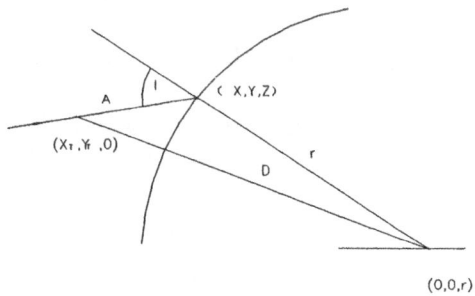

Figure 6.2. Determination of the physical meaning for Eq. (6.12)

From Fig.6.2 it can be seen that all the lines lie in the plane of incidence and by using the cosine law it can be stated that:

$$D^2 = X_T^2 + Y_T^2 + r^2 = A^2 + r^2 + 2 \cdot A \cdot r \cdot \cos I \tag{6.13}$$

Solving Eq. (6.13) for $\cos I$ is obtained:

$$n_{j-1} \cdot \cos I = \frac{H - c_j \cdot n_{j-1}^2 \cdot \left(\dfrac{A}{n_{j-1}}\right)^2}{2 \cdot \left(\dfrac{A}{n_{j-1}}\right)} \tag{6.14}$$

Substituting the expression for $\dfrac{A}{n_{j-1}}$ with the negative sign from the Eq. (6.12) gives:

$$n_{j-1} \cdot \cos I = n_{j-1} \cdot \sqrt{\left(\frac{B}{n_{j-1}}\right)^2 - c_j \cdot H} \tag{6.15}$$

Returning Eq. (6.15) to Eq. (6.12) and after necessary transformations gives:

$$\frac{A}{n_{j-1}} = \frac{H}{B + n_{j-1} \cdot \cos I} \tag{6.16}$$

6.4.3   Ray refraction procedure at the spherical surface

The ray refraction on the spherical surface is based on the Snellius – Descartes law which states that the vector of the incident ray, the refracted ray and the normal at the surface lies in one plane. In Fig.6.3 this plane is shown.

The vector $\vec{S}_0$ is the vector of the incident ray, the vector $\vec{S}_1$ is the vector of the refracted ray and the third side of the triangle is parallel to the normal to the surface i.e. parallel to the surface radius and has the length Γ. The unit vector is quotient of the vector parallel to the normal divided by the radius r.

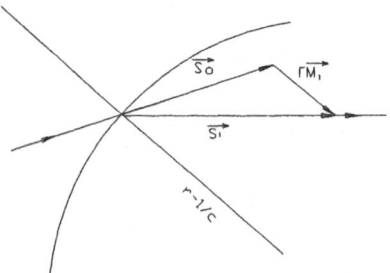

Figure 6.3. Triangle for the law of refraction

Hence

$$\vec{M} = c_j \cdot \left[(0 - X_j) \cdot \vec{i} + (0 - Y_j) \cdot \vec{j} + (r_j - Z_j) \cdot \vec{k} \right]$$
$$= c_j \cdot \left[(-X_j) \cdot \vec{i} + (-Y_j) \cdot \vec{j} + (r_j - Z_j) \cdot \vec{k} \right] \tag{6.17}$$

where $\vec{i}, \vec{j}, \vec{k}$ are unit vectors along the coordinate axes. From Fig. 6.3 it can be seen that:

$$\vec{s}_1 - \vec{s}_0 = -c_j \cdot X_j \cdot \Gamma \cdot \vec{i} - c_j \cdot Y_j \cdot \Gamma \cdot \vec{j} + c_j \cdot \Gamma \cdot (r_j - Z_j) \cdot \vec{k} \tag{6.18}$$

The equations for the $\vec{s}_0, \vec{s}_1$ are:

$$\vec{s}_0 = n_{j-1} \cdot \vec{Q}_0 = K_{j-1} \cdot \vec{i} + L_{j-1} \cdot \vec{j} + M_{j-1} \cdot \vec{k}$$
$$\vec{s}_1 = n_j \cdot \vec{Q}_1 = K_j \cdot \vec{i} + L_j \cdot \vec{j} + M_j \cdot \vec{k} \tag{6.19}$$

Subtracting the vector $\vec{s}_0$ from the vector $\vec{s}_1$ it is obtained:

$$\vec{s}_1 - \vec{s}_0 = (K_j - K_{j-1}) \cdot \vec{i} + (L_j - L_{j-1}) \cdot \vec{j} + (M_j - M_{j-1}) \cdot \vec{k} \tag{6.20}$$

The relations between the old and the new optical direction cosines are obtained from Eq. (6.20) and Eq. (6.18):

$$(K_j - K_{j-1}) \cdot \vec{i} + (L_j - L_{j-1}) \cdot \vec{j} + (M_j - M_{j-1}) \cdot \vec{k} =$$
$$-c_j \cdot X_j \cdot \Gamma \cdot \vec{i} - c_j \cdot Y_j \cdot \Gamma \cdot \vec{j} + c_j \cdot \Gamma \cdot (r_j - Z_j) \cdot \vec{k} \tag{6.21}$$

or in scalar form:

$$K_j = K_{j-1} - c_j \cdot X_j \cdot \Gamma$$
$$L_j = L_{j-1} - c_j \cdot Y_j \cdot \Gamma$$
$$M_j = M_{j-1} - (c_j \cdot Z_j - 1) \cdot \Gamma \tag{6.22}$$

Only the constant Γ is not yet defined. From Fig. 6.3 it can be seen that the constant Γ is equal to the difference of the $\vec{s}_0$ and $\vec{s}_1$ vector projections on the direction of the radius.

$$\Gamma = n_j \cdot \cos I' - n_{j-1} \cdot \cos I \tag{6.23}$$

Snellius-Descartes law in the scalar form is:

$$n_j \cdot \sin I' = n_{j-1} \cdot \sin I \tag{6.24}$$

or after required trigonometrically transformations:

$$n_j \cdot \cos I' = n_j \cdot \sqrt{\left(\frac{n_{j-1}}{n_j} \cdot \cos I\right)^2 - \left(\frac{n_{j-1}}{n_j}\right)^2 + 1} \tag{6.25}$$

With Eq. (6.25) all the necessary equations for calculating the ray trace are shown.

6.4.4   Ray trace equations summary

In this section all ray trace equations will be written in the order of use.

$$\frac{d_{j-1}}{n_{j-1}} = \left(t_{j-1} - z_{j-1}\right) \cdot \frac{1}{M_{j-1}}$$

$$X_T = X_{j-1} + \Delta X = X_{j-1} + d_{j-1} \cdot \frac{K_{j-1}}{n_{j-1}}$$

$$Y_T = Y_{j-1} + \Delta Y = Y_{j-1} + d_{j-1} \cdot \frac{L_{j-1}}{n_{j-1}}$$

$$B = M_{j-1} - c_j \cdot \left(X_T \cdot K_{j-1} + Y_T \cdot L_{j-1}\right)$$

$$H = c_j \cdot \left(X_T^2 + Y_T^2\right)$$

$$n_{j-1} \cdot \cos I = n_{j-1} \cdot \sqrt{\left(\frac{B}{n_{j-1}}\right)^2 - c_j \cdot H}$$

$$\frac{A}{n_{j-1}} = \frac{H}{B + n_{j-1} \cdot \cos I}$$

$$X_j = X_T + \frac{A}{n_{j-1}} \cdot K_{j-1}$$

$$Y_j = Y_T + \frac{A}{n_{j-1}} \cdot L_{j-1}$$

$$Z_j = \frac{A}{n_{j-1}} \cdot M_{j-1}$$

$$n_j \cdot \cos I' = n_j \cdot \sqrt{\left(\frac{n_{j-1}}{n_j} \cdot \cos I\right)^2 - \left(\frac{n_{j-1}}{n_j}\right)^2 + 1}$$

$$\Gamma = n_j \cdot \cos I' - n_{j-1} \cdot \cos I$$

$$K_j = K_{j-1} - c_j \cdot X_j \cdot \Gamma$$

$$L_j = L_{j-1} - c_j \cdot Y_j \cdot \Gamma$$

$$M_j = M_{j-1} - \left(c_j \cdot Z_j - 1\right) \cdot \Gamma$$

Chapter 7

Aberrations

The aberrations can be defined as any systematic deviation from an ideal path of the image forming rays passing through an optical system, which causes the image to be imperfect. In other words the aberrations are differences from the real and the ideal image. The ideal image is formed under assumption that all rays, emerging from the one point on the object, after traversing the optical system, pass through one point on the image. The real image is calculated by the ray trace through the optical system. For the paraxial rays (rays that are close to the optical axis) the aberrations are small and can be neglected. But for all other rays that are passing on the finite distance from the optical axis and with the finite field angle the aberrations becomes considerable because they distort image.

The basic reason for the aberrations is in the fact that the lenses are formed from the spherical surfaces, which do not refract rays at the same way as it is assumed in the paraxial approximation. These aberrations are called the geometrical aberrations. Another reason for the aberrations is connected with the light dispersion. Since the index of refraction for the optical glass is the function of the light wavelength then the effective focal length and the other optical system characteristics are also a function of the wavelength. Therefore the rays emitted from the same object point and having different wavelengths after passing through the optical system do not converge in the same image point although each of them alone can be ideal. This kind of aberrations is called the chromatic aberrations.

For the purpose of optical design aberrations are broadly divided into two large groups: the monochromatic and the chromatic aberrations. The monochromatic aberrations or the geometrical aberrations are aberrations occuring when the optical system is illuminated by monochromatic light. The chromatic aberrations occur due to variations in optical system properties with wavelength of the incident polychromatic radiation. In practice, aberrations occur in combinations rather than alone. The system of classifying aberrations makes the analysis much simpler and gives a good description of optical system image quality.

Normally aberrations are measured by the amount by which rays miss the paraxial image point. The aberrations are measured by the linear displacement of the points at which real rays intersect the image surface, which is defined as image plane for the perfect optical system.

It should be kept in mind that aberrations are unavoidable in actual optical systems and one of the aims of optical design is to correct optical system being designed for aberrations. Some aberrations, however, inevitably remain and the purpose of the image quality evaluations is to determine how large the residual aberrations are. In general a modern optical design takes aberrations into account by solving following problems:

- determine the residual aberrations for an optical system with specified constructional parameters;

- evaluate the variable constructional parameters for an optical system which will keep the residual aberrations within the specified tolerable values. These tolerable values can be derived from the purpose of the optical system.

In this chapter only description of the basic monochromatic and chromatic aberrations and equations for their calculation will be given. All the theory of aberrations such as the wave aberrations, the third (Seidel) aberrational theory and the fifth (Buchdahl) aberrational theory are omitted.

7.1   Spherical aberration

The general definition of the spherical aberration can be the change of the focal length with the change of aperture. The spherical aberration is defined by the MIL – STD – 1241A (Optical terms and definitions) as a symmetrical optical defect of lenses and spherical mirrors in which light rays that come from a common axial point and enter the optical system at different distances from the optical axis after exiting the optical system do not come to the paraxial focus but form a circular blur. Main cause for the spherical aberration is that the angle between the incident ray (which is parallel with the optical axis) and the normal to the spherical surface is not the same at different heights. This angle is growing and therefore the rays on the larger distances from the optical axis during the refraction through the optical surface are turned with greater angle of refraction toward the optical axis and because of that cross the optical axis before the rays on the lower heights. For the focal systems the longitudinal and the transverse spherical aberration is usually calculated and expressed in linear units. For afocal systems only one spherical aberration is usually calculated and expressed in the angular units.

7.1.1   Longitudinal spherical aberration of the focal systems

The longitudinal spherical aberration is defined as the distance measured along the optical axis between the intersection point of the axial ray for the referent (basic) wavelength and the intersection point of the paraxial ray (the paraxial focus). The referent wavelength is usually the middle wavelength for the given spectral range. For the visible light it is usually the Fraunhofer's d spectral line, which represents yellow helium light ($\lambda_d = 587.56$ nm).

$$\delta s' = s' - s_0'$$
(7.1)

where:

$\delta s'$ is the longitudinal spherical aberration;

s' is the back focal length calculated by the axial ray trace;

s_0' is the back focal length calculated by the paraxial ray trace.

The graphical representation of the longitudinal spherical aberration is shown in Fig. 7.1.

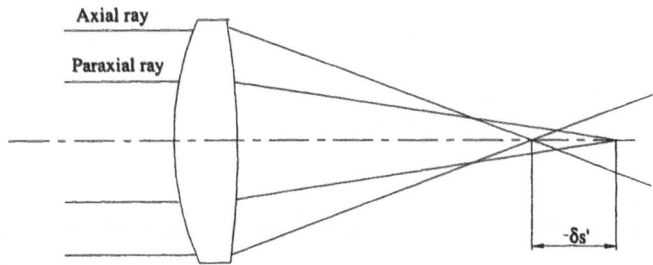

Figure 7.1. The longitudinal spherical aberration

The equation for the calculation of the back focal length by the axial ray trace is:

$$s' = Z' - Y' \cdot \frac{M'}{L'} \tag{7.2}$$

where:

Z' is the z coordinate (along the optical axis) of the exit ray;

Y' is the y coordinate of the exit ray;

M' is the direction cosines of the exit ray along the z axis;

L' is the direction cosines of the exit ray along the y axis.

The equations for the calculation of the back focal length by the paraxial rays are:

The basic equation of the paraxial optics – the ray trace at the refraction surface is:

$$s_j' = \frac{n_{j+1}}{\dfrac{n_j}{s_j} + \dfrac{\left(n_{j+1} - n_j\right)}{r_j}} \tag{7.3}$$

where:

s_j' is the distance of the refracted ray from the jth surface of the optical system

s_j is the distance of the incident ray from the jth surface of the optical system

r_j is the radius of the curvature for the jth surface of the optical system;

n_j is the index of refraction for the optical medium before the jth surface of the optical system;

n_{j+1} is the index of refraction for the optical medium before the (j+1)th surface of the optical system.

The transfer equation from one refraction surface to another surface is:

$$s_{j+1} = s_j' - d_j \tag{7.4}$$

where:

s_{j+1} is the distance of the incident ray from the (j+1)th surface of the optical system;

d_j is the distance between the jth and the (j+1)th surface of the optical system.

The longitudinal spherochromatism or the cromatic variation of the longitudinal spherical aberration is the longitudinal spherical aberration for the lowest and the highest wavelength. For the visible light the minimum wavelength is usually the Fraunhofer's F spectral line representing the blue hydrogen light ($\lambda_F = 486.13$ nm) and the maximum wavelength is usually the Fraunhofer's C spectral line representing the red hydrogen light ($\lambda_C = 656.27$ nm).

$$\begin{aligned} \delta s_{min}' &= s_{min}' - s_0' \\ \delta s_{max}' &= s_{max}' - s_0' \end{aligned} \tag{7.5}$$

where:

$\delta s_{min}', \delta s_{max}'$ are the longitudinal spherochromatism for the lowest and the highest wavelength;

s_{min}', s_{max}' are the back focal length calculated by the axial ray trace for the lowest and the highest wavelength.

The axial chromatic aberration is defined as the difference between the longitudinal spherochomatism for the lowest and the highest wavelength:

$$\Delta s' = \delta s_{min}' - \delta s_{max}' \tag{7.6}$$

where $\Delta s'$ is the axial chromatic aberration.

7.1.2   Transverse spherical aberration of the focal systems

The transverse spherical aberration is defined as the vertical distance from the optical axis to the intersection of the axial ray with the paraxial image plane.

$$\delta l' = \delta s' \cdot \frac{L'}{M'} \tag{7.7}$$

where $\delta l'$ is the transverse spherical aberration.

The graphical representation of the transverse spherical aberration is shown in Fig. 7.2.

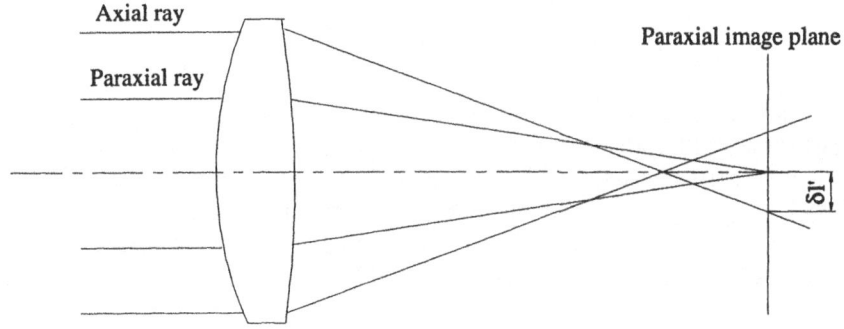

Figure 7.2. The transverse spherical aberration

The transverse spherochromatism or the chromatic variation of the transverse spherical aberration is the transverse spherical aberration for the lowest and the highest wavelength. For the visible light minimum wavelength is usually the Fraunhofer's F spectral line which represents the blue hydrogen light ($\lambda_F = 486.13$ nm) and the maximum wavelength is usually the Fraunhofer's C spectral line representing the red hydrogen light ($\lambda_C = 656.27$ nm).

The equations for the transverse spherochromatism or the chromatic variation of the transverse spherical aberration are:

$$
\begin{aligned}
\delta l'_{min} &= \delta s'_{min} \cdot \frac{L'_{min}}{M'_{min}} \\
\delta l'_{max} &= \delta s'_{max} \cdot \frac{L'_{max}}{M'_{max}}
\end{aligned}
\tag{7.8}
$$

where:

$\delta l'_{min}, \delta l'_{max}$ are the transverse spherochromatism for the lowest and the highest wavelength;

L'_{min}, L'_{max} are the direction cosines of the exit ray along the y axis for the lowest and the highest wavelength;

M'_{min}, M'_{max} are the direction cosines of the exit ray along the z axis for the lowest and the highest wavelength.

The transverse chromatic aberration is defined as the difference between the transverse spherochromatism for the lowest and the highest wavelength:

$$\Delta l' = \delta l'_{min} - \delta l'_{max} \qquad (7.9)$$

where $\Delta l'$ is the transverse chromatic aberration.

7.1.3   Spherical aberration of the afocal systems

The spherical aberration of the afocal systems is calculated as the exit angle from the last surface of the afocal system for the referent (basic) wavelength. The referent wavelength is usually the middle wavelength for the given spectral range. For the visible light it is usually the Fraunhofer's d spectral line, which represents the yellow helium light ($\lambda_d = 587.56\,\text{nm}$). This exit angle can be calculated when the direction cosine for the y-axis from the last surface of the afocal system is expressed in degrees:

$$\delta u' = u'_d \qquad (7.10)$$

where $\delta u'$ is the spherical aberration of the afocal systems and u'_d is the exit angle from the last surface of the afocal system for the Fraunhofer's d spectral line expressed in degrees.

The spherochromatism or the chromatic variation of the spherical aberration is the spherical aberration for the lowest and the highest wavelength. For the visible light the minimum wavelength is usually the Fraunhofer's F spectral line which represents the blue hydrogen light ($\lambda_F = 486.13\,\text{nm}$) and the maximum wavelength is usually the Fraunhofer's C spectral line representing the red hydrogen light ($\lambda_C = 656.27\,\text{nm}$).

$$\begin{aligned}
\delta u'_{min} &= u'_{min} \\
\delta u'_{max} &= u'_{max}
\end{aligned} \qquad (7.11)$$

where:

$\delta u'_{min}, \delta u'_{max}$ are the spherochromatism for the lowest and the highest wavelength;

u'_{min}, u'_{max} are the exit angle from the last surface of the afocal system for the lowest and the highest wavelength.

The chromatic aberration is defined as the difference between the spherochromatism for the lowest and the highest wavelength:

$$\Delta u' = \delta u'_{min} - \delta u'_{max} \tag{7.12}$$

where $\Delta u'$ is the chromatic aberration.

The optical system that have corrected the spherical aberration and the chomatic aberration are called the achromatic optical systems. The spherical aberration is the basic aberration that must be corrected for the every optical system. In other words the spherical aberration must be corrected first and if there is more variable constructional parameters then other aberrations may be corrected. The simplest optical system with the spherical surfaces that can be freed from the spherical aberration is the achromatic doublet which consists of the positive and negative lens.

The chromatic aberration is corrected by the appropriate glass selection. For the correction of the chromatic aberration it is necessary to have the the optical system with minimum two elements: one made of the crown glass and the other made of the flint glass. So the achromatic doublet is the simplest optical system that can be corrected for the spherical aberration and chromatic aberration.

7.2 Sine condition and isoplanetism

If the optical system produces an aberration free image of the point on the axis then to produce an aberration free image of an infinitesimal line segment perpendicular to the optical axis the system must satisfy the sine condition which is defined by the following equation:

$$n' \cdot dy' \cdot \sin\sigma' = n \cdot dy \cdot \sin\sigma \tag{7.13}$$

where:

dy, dy' are the infinitesimal object and image line segments perpendicular to the optical axis;

σ, σ' are the slopes of the rays through the axial points of the object and the image;

n, n' are the refractive index of the object and image media.

This condition must be satisfied for any values of the angle σ. In theory Eq. (7.13) is known as the Lagrange–Helmholtz invariant.

For the infinitely distant object point the sine condition becomes:

$$\frac{H}{\sin\sigma'} = f' = f'_0 \tag{7.14}$$

where:

f' is the effective focal length calculated by the axial ray trace;

f'_0 is the effective focal length calculated by the paraxial ray trace;

H   is the incidence height at the entrance pupil for an axial ray i.e. the ray that enters the optical system parallel to the optical axis and emanates from the optical system at an angle σ' with the optical axis;

$\sin \sigma'$ is the angular aperture in the image space;

σ' is the angle of the exit ray with the optical axis.

In order to fulfil the sine condition Eq. (7.14) must be satisfied for any value of the incident ray height from $H = 0$ to $H = H_{max}$ where the maximal height H_{max} is equal to the half of the enterance pupil height.

Because the ultimate value of the angular aperture in the image space is 90° the maximum aperture ratio of the optical system satisfying the sine condition is confined by the inequality $D/f' < 1:0.5$ (i.e. the relative aperture $f'/D > f'/0.5$).

The conugate axial points for which spherical aberration is virtually absent and the sine condition is met are referred to as aplanatic. Optical systems capable of meeting these conditions include microlenses. In many cases, however, optical systems cannot produce a perfect image for an axial point. The large pupil optical systems are corrected for spherical aberration for two, seldom for three, rays; the other rays of the axial ray bundle exibit unremovable spherical aberration.

Usually, real optical systems do not fulfil the sine condition, so the deviation from the sine condition is calculated as the one of the aberrations.

The deviation from the sine condition for an infinitely distant object is:

$$\delta f' = \frac{H}{\sin \sigma'} - f_0'$$ (7.15)

where $\delta f'$ is the deviation from the sine condition for an infinitely distant object.

The deviation from the sine condition for an object at a finite distance from the optical system is:

$$\delta \beta = \beta - \beta_0 = \frac{n \cdot \sin \sigma}{n' \cdot \sin \sigma'} - \frac{n \cdot \sin \sigma_0}{n' \cdot \sin \sigma_0'}$$ (7.16)

where:

$\delta\beta$ is the deviation from the sine condition for an object at a finite distance from the optical system;

β is the transverse magnification for axial rays;

β_0 is the transverse magnification for paraxial rays;

σ, σ' are the incident and the exit angle for the axial ray trace;

σ_0, σ_0' are the incident and the exit angle for the paraxial ray trace.

At real optical systems with the residual spherical aberration it is gravitated toward the fulfilment of the isoplanetism condition representing the generalized

Abe's sine condition. The isoplanetism condition is fulfilled when the image quality for the points near the optical axis is the same as for the axial point. The isoplanetism condition is not so strict as the sine condition.

The deviation from the isoplanetism for an object at a finite distance from the optical system is:

$$\eta = \frac{\delta\beta}{\beta_0} - \frac{\delta s'}{s' - t'} \qquad (7.17)$$

where:

η is the deviation from the isoplanetism;

$s' - t'$ is the distance from the exit pupil plane to the image plane;

s' is the distance from the last surface of the optical system to the image plane;

t' is the distance from the last surface of the optical system to the exit pupil plane.

The deviation from the isoplanetism for an infinitely distant object is:

$$\eta = \frac{\delta f'}{f_0'} - \frac{\delta s'}{s' - t'} \qquad (7.18)$$

7.3 Astigmatism

The astigmatism is a phenomenon where the rays from one pencil of rays, viewed in two mutual normal planes (the meridional and the sagital) after refraction through the optical system, do not intersect in the one point nor they intersect in the paraxial image plane but they form an image – the meridional and the sagital focal line which are located at a certain distance from the paraxial image plane and under the 90° angle from which one focal line is horizontal and the other is vertical.

The meridional plane can be defined as a plane set through the object point and containing the optical axis. The sagital plane is a plane set through the object point and normal to the meridional plane.

In order to fully understand the astigmatism it is necessary to follow the image forming from the off-axis point B for the following images:

– the paraxial image in the point B',

– the meridional image in the point B_m', more exactly the meridional focal line M_1M_2,

– the sagital image in the point B_s', more exactly the sagital focal line S_1S_2,

– the best image in the point B_a', because there is the circle of least confusion.

The graphical representation of the astigmatism is shown in Fig. 7.3:

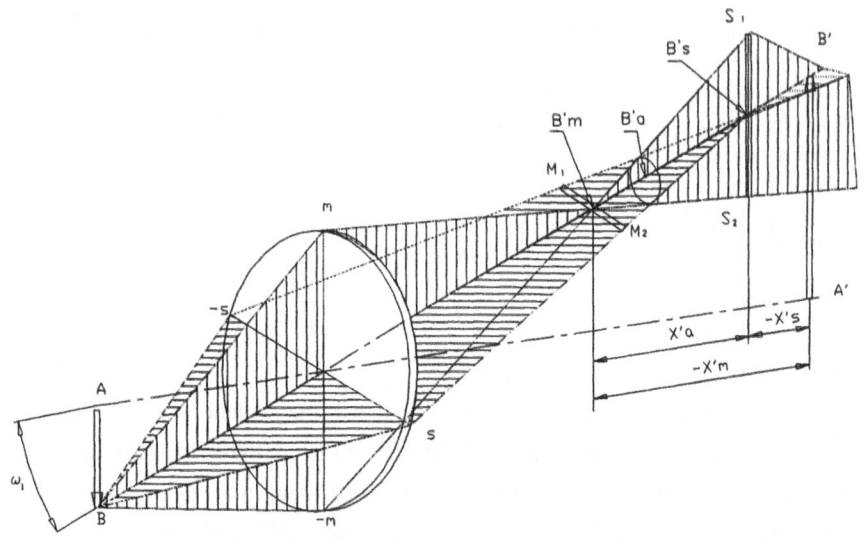

Figure 7.3. The Astigmatism

The following labels are used in Fig. 7.3:

AA′ is the optical axis;

BB′ is the principal ray forming the field angle ω with the optical axis;

mm is the meridional plane;

ss is the sagital plane;

A′ is the image of the axial point *A* calculated by the paraxial ray trace;

B′ is the image of the off-axis point *B* calculated by the paraxial ray trace;

B'_m is the image of the off-axis point *B* calculated by the elementary astigmatic ray trace in the meridional plane, that is the point B'_m is the point of intersection of all rays travelling in the meridional plane with the principal ray;

B'_s is the image of the off-axis point *B* calculated by the elementary astigmatic ray trace in the sagital plane, that is the point B'_s is the point of intersection of all rays travelling in the sagital plane with the principal ray;

M_1M_2 is the meridional focal line, the image of the off-axis point *B* calculated by the elementary astigmatic ray trace in the sagital plane, that is the rays travelling in the sagital plane and they are not converged enough to collect in the point B'_m but they are distributed along the segment M_1M_2 as they pass through the plane which contains the point B'_m;

S_1S_2 is the sagital focal line, the image of the off-axis point *B* calculated by the elementary astigmatic ray trace in the meridional plane, that is the rays travelling in the meridional plane and because they are already converged in the point B'_m now they diverge and they are distributed along the segment $S_1 S_2$ as they pass through the plane which contains the point B'_s ;

B'_a is the point, more exactly the plane of best image for the off-axis point B . Rays that are travelling in the meridional and the sagital plane form in the best image plane the circular spot called circle of least confusion;

$-x'_s$ is the distance between the sagital focal line and the paraxial image plane;

$-x'_m$ is the distance between the meridional focal line and the paraxial image plane;

x'_a is the distance between the sagital and the meridional focal line.

The principal ray from any axial infinitesimal narrow ray bundle passes through the centre of the optical surface curvature so the size of curvature radius in the meridional (mm) and the sagital (ss) plane is equal ($r_m = r_s$).

For the infinitesimal narrow ray bundles that are starting from the off-axis point B the conditions of travelling in the meridional and sagital plane will be different. The principal ray, around which other rays are symmetrically distributed, do not pass in general through the centre of the optical surface curvature so the size of curvature radius in the meridional (mm) and the sagital (ss) plane will be different ($r_m \neq r_s$). The exit wave form that correspondents to the elementary extra-axial ray bundle is not any more spherical but deformed. Rays travelling in the meridional and the sagital plane intersect with the principal ray in points B'_m and B'_s in the image space at the distances $-x'_m$ and $-x'_s$ from the paraxial image plane B' .

All rays starting from the off-axis object point B and travelling in the meridional plane converge to the point B'_m , while rays starting from the same point B and travelling in the sagital plane form the horizontal segment instead of the point as expected by paraxial theory. The same is true for the plane that contains the point B'_s in which converge all rays starting from the off-axis object point B and travelling in the sagital plane. The rays starting from the same point B and travelling in the meridional plane form the vertical segment.

This phenomenon, more exactly the optical system defect where the image of the off-axis point is formed as the two mutually normal lines contained in planes on the different distances from the paraxial image plane, is called astigmatism.

The astigmatism of the optical system is calculated by the tracing the rays infinitesimally close to the principal ray in the meridional and the sagital plane. The Young – Abbe equations for the elementary astigmatic ray in the meridional and the sagital plane are used for ray trace calculation:

– for the meridional plane:

$$\frac{n' \cdot \cos^2 \varepsilon'}{t'_m} - \frac{n \cdot \cos^2 \varepsilon}{t_m} = \frac{n' \cdot \cos \varepsilon' - n \cdot \cos \varepsilon}{r} \tag{7.19}$$

– for the sagital plane:

$$\frac{n'}{t'_s} - \frac{n}{t_s} = \frac{n' \cdot \cos\varepsilon' - n \cdot \cos\varepsilon}{r} \tag{7.20}$$

where:

t_m is the distance in the object space along the elementary astigmatic ray in the meridional plane from the starting point to the point on the optical surface;

t'_m is the distance in the image space along the elementary astigmatic ray in the meridional plane from the point on the optical surface to the point which represents the image of the starting point;

n is the index of refraction for the optical medium before the optical surface;

n' is the index of refraction for the optical medium after the optical surface;

r the radius of curvature for the optical surface;

ε is the angle between the elementary astigmatic ray and the normal on the optical surface in the object space;

ε' is the angle between the elementary astigmatic ray and the normal on the optical surface in the image space;

t_s is the distance in the object space along the elementary astigmatic ray in the sagital plane from the starting point to the point on the optical surface;

t'_s is the distance in the image space along the elementary astigmatic ray in the sagital plane from the point on the optical surface to the point which represents image of the starting point.

The equations for the ray transfer from one optical surface to another are:

— for the meridional plane:

$$t'_m = t_m - \tilde{d}_k \tag{7.21}$$

— for the sagital plane:

$$t'_s = t_s - \tilde{d}_k \tag{7.22}$$

where $\tilde{d}_k$ is the distance between two optical surfaces along the elementary astigmatic ray.

It is necessary to solve Eq. (7.19) and (7.20) with respect to the variables t'_m and t'_s in order to use them on computer:

— for the meridional plane:

$$t'_m = \frac{\cos^2 \varepsilon}{\dfrac{\left[\left(\dfrac{n}{n'}\right) \cdot \cos^2 \varepsilon'\right]}{t_m} + \dfrac{\left[\cos \varepsilon' - \left(\dfrac{n}{n'}\right) \cdot \cos \varepsilon\right]}{r}} \qquad (7.23)$$

– for the sagital plane:

$$t'_s = \frac{1}{\dfrac{\left(\dfrac{n}{n'}\right)}{t_s} + \dfrac{\left[\cos \varepsilon' - \left(\dfrac{n}{n'}\right) \cdot \cos \varepsilon\right]}{r}} \qquad (7.24)$$

The projections of the meridional and the sagital focal line on the optical axis are necessary to be defined for the calculation of astigmatism for the optical systems.

– the projection of the meridional focal line on the optical axis is:

$$x'_m = t'_m \cdot \cos \varepsilon' + x'_t - s'_0 \qquad (7.25)$$

– the projection of the sagital focal line on the optical axis is:

$$x'_s = t'_s \cdot \cos \varepsilon' + x'_t - s'_0 \qquad (7.26)$$

where:

t'_m is the distance along the elementary astigmatic ray in the meridional plane calculated by the elementary astigmatic ray trace through the whole optical system;

ε' is the refraction angle of the ray exiting optical system calculated by the principle ray trace;

x'_t is the x coordinate of the principle ray intersection point with the last optical surface;

s'_0 is the back focal length calculated by the paraxial ray trace;

t'_s is the distance along the elementary astigmatic ray in the sagital plane calculated by the elementary astigmatic ray trace through the whole optical system.

The astigmatism is defined as:

– for the focal systems:

$$x'_a = x'_s - x'_m \qquad (7.27)$$

− for the afocal systems:

$$A = 1000 \cdot \left(\frac{1}{t'_s} - \frac{1}{t'_m} \right)$$
(7.28)

For the afocal systems the astigmatism is calculated in dioptres.

7.4   Field curvature

The field curvature is closely connected with the astigmatism and they are always discussed together. From Eq. (7.25) and (7.26) can be seen that the position of the meridional and the sagital focal line depends on the field angle. They are formed on the different distances from the paraxial image plane. If all places of forming focal lines connected together and positioned in the meridional plane two curves representing change of the meridional and the sagital focal line positions with the change of the field angle are calculated. The curve that is calculated as the arithmetical mean of these two curves is called the field curvature.

The physical essence of the aberration field curvature is that the flat image after projection through the optical system will not be any more flat but will be curved i.e. if the object is positioned in the plane after the projection through the optical system image will not be in the plane but in the space and it will form paraboloid.

The field curvature for the focal systems can be defined as the arithmetical mean of the meridional and the sagital focal lines projections on the optical axis:

$$x'_k = \frac{x'_s + x'_m}{2}$$
(7.29)

The field curvature for the afocal systems is:

$$Kr = 500 \cdot \left(\frac{1}{t'_m} + \frac{1}{t'_s} \right)$$
(7.30)

For the afocal systems the field curvature is calculated in dioptres.

To ensure sharp imagery over the entire field, the optical system should be corrected for both astigmatism and field curvature. The optical systems corrected for both aberrations within a certain angular field, with the residual aberrations having the tolerable values over the entire field of view are called anastigmats. The simplest optical system with spherical surfaces that can be called anastigmat (i.e. corrected for the spherical aberaction, the chromatism, the coma, the astismatism and the field curvature) is the Cooke triplet which consists three separated lenses two positive and one negative between the positive ones.

7.5 Distortion

The distortion in regard to all other aberrations has special features. It does not cause the vagueness of the object image. If in the optical system all other aberrations are corrected and only pure distortion is left, then the images of all the object points will be also clear points. The distortion causes the deformation of the object image in the geometrical sense. In other words the distortion is an aberration appearing as the curvature of straight lines in the object, i.e. as a breakdown of the geometric similarity between an object and an image.

The distortion arises from the incomplete realization of the well known geometrical optics law that the transverse magnification for the pair of conjugated and normal to the optical axis planes is constant. Nonfeasance of this law leads to the deformation of the image. In geometrical optics the transverse magnification of an optical system is defined as:

$$\beta = \frac{y'}{y} \tag{7.31}$$

where:

y' is the height of the image;

y is the height of the object.

If the transverse magnification remains constant for any height y and equal to the transverse magnification of the perfect optical system then the distortion is absent. The optical system free from distortion is called ortoscopic.

The graphical representation of distortion is given in Fig. 7.4.

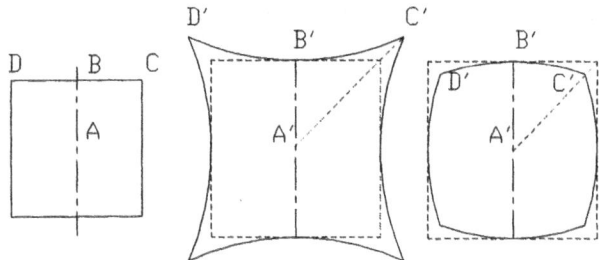

Figure 7.4. The distortion

In Fig. 7.4 two images of the square object are represented. The first image is obtained when the transverse magnification increases while moving away from the optical axis, then the distortion also increases. This type of distortion is called the "positive" or the pin-cushion distortion. The second image is obtained when the transverse magnification decreases while moving away from the optical axis, then the distortion also decreases. This type of distortion is called the "negative" or the barrel distortion.

The distortion for the focal systems is calculated as the difference between the real image, which is formed by the principal ray, and the paraxial image. The distortion is usually computed in percentage:

$$\delta l' = \frac{l' - l_0'}{l_0'} \cdot 100 \qquad (7.32)$$

where:

l' is the size of the real image formed by the principal ray;

l_0' is the size of the paraxial image.

If the object is in infinity the size of the paraxial image is calculated as:

$$l_0' = f' \cdot \text{tg}\omega \qquad (7.33)$$

where ω is the field angle.

If the object is at finite distance from an optical system the size of the paraxial image is calculated as:

$$l_0' = f' \cdot \frac{-y}{t - s} \qquad (7.34)$$

where:

y the object height;

s is the distance from the object to the first surface of the optical system;

$t - s$ is the distance between the object and the entrance pupil.

The distortion for the afocal systems is calculated as:

$$D = \frac{\text{tg}\omega' - \text{tg}\omega_0'}{\text{tg}\omega_0'} \cdot 100 \qquad (7.35)$$

where:

ω' is the apparent field angle calculated by the principle ray trace;

ω_0' is the apparent field angle calculated by the paraxial ray trace.

7.6 Coma

The coma appears for the wide ray bundles entering an optical system at a certain angle to the optical axis. The coma is the aberration which essentially spoils the symmetry of the ray bundle, which after passing through the optical system is no longer symmetric in relation to the principal ray. The corrupted symmetry for the ray exiting the optical system is explained by unequal refracting conditions for rays entering the optical system in different zones of the entrance pupil. As the result of this asymmetry the spot, which is produced by the dispersion of the rays starting from the extra-axial object point and passing through the optical system, loses its circular form characteristic for the spherical aberration and obtains the form of the comet with tail.

The coma for the focal systems can be presented as the change of the transverse magnification i.e. the change of the image size with the change of the aperture. The coma can be defined as the vertical distance from the principal ray to the upper or the lower oblique ray:

$$K = \frac{l'_u + l'_l}{2} - l'_p \qquad (7.36)$$

where:

l'_u and l'_l are the image size calculated by the upper or the lower oblique ray trace;

l'_p is the image size calculated by the principal ray trace.

The coma for the afocal systems can be defined as the angle covering the principal ray and the upper or the lower oblique ray:

$$K = \frac{\omega'_u + \omega'_l}{2} - \omega'_p \qquad (7.37)$$

where:

ω'_u and ω'_l are the apparent field angle calculated by the upper or the lower oblique ray trace;

ω'_p is the apparent field angle calculated by the principle ray trace.

Chapter 8

Damped least squares optimization implementation

8.1   General description of the damped least squares optimization

Part III of this monography will give a detailed description of different optimization methods used for the optical system optimization. The following optimization methods will be described:

– the damped least squares (DLS) method as a representative of the classical optimization methods;

– the adaptive steady-state genetic algorithm;

– the two membered evolution strategy – EVOL;

– the multimembered evolution strategies – GRUP, REKO and KORR.

A detailed mathematical theory for the classical optimization methods is given in Chapter 2, for the genetic algorithms in Chapter 3 and for the evolution strategies in Chapter 4. The mathematical theory of the ray trace is given in Chapter 6 and the aberrational theory in Chapter 7. So in Chapters 2 to 7 all necessary mathematical theory needed for understanding the optimization methods and their implementation is given. This and the next four chapters will give a detailed description of the optimization methods implementation. For each optimization method the flow chart will be given and the main parts of the optimization algorithm will be described in details.

The basic flow chart for the DLS optimization method is given in Fig. 8.1. The DLS optimization starts with the ray trace through the optical system and definition of the weighting factors for the aberrations forming the merit function. The detailed description of the merit function and its construction from the aberrations and the weighting factors is given in Section 8.2. The optical designer can accept default values for the weighting factors, which simplifies necessary input data for the optimization method.

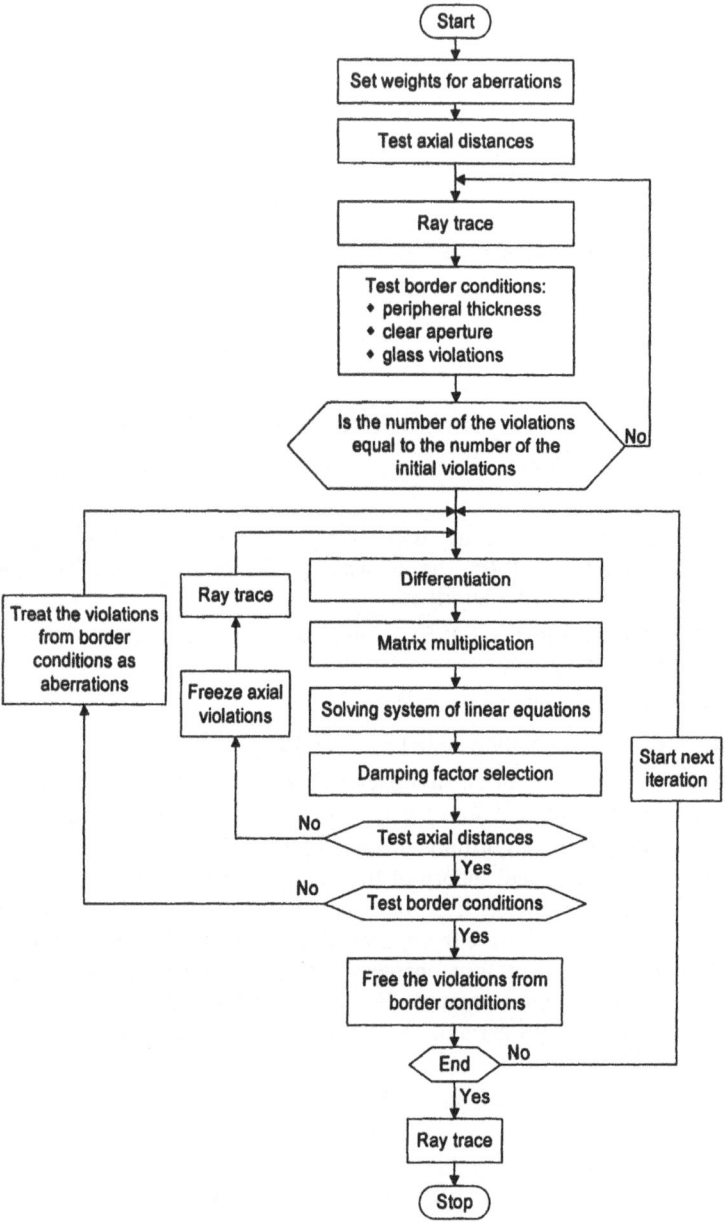

Figure 8.1. The basic flow chart for the DLS optimization

The default values for the weighting factors are usually the best values for the standard types of optical systems. By using default values, it is possible to obtain sufficiently good optical system designs relatively easy and quickly. The author recommends that the optical system designer always use the default values for the weighting factors first and see the results. If the results are satisfactory then there is no need to change anything. If the results are not satisfactory then the optical system designer, after studying the optimization results and according to his experience and intuition, can change the values for the weighting factors. The experienced optical designer with the special requirements that must be fulfilled has the opportunity to modify all weighting factors completely and to adapt them to his needs.

After defining the weighting factors for the aberrations and calculating the merit function for the starting optical system, the boundary conditions are checked. First, all the axial distances (the lens thickness or the distances between optical elements) must be greater than the minimum allowed distance. If any distance is smaller than the minimum allowed distance, then this distance is assigned a value for the minimum allowed distance and it is not allowed to change this parameter in the optimization. After checking the axial distances the rest of the boundary conditions are tested. The following boundary conditions are tested:

- the peripheral thickness for the convex lenses which ought to be greater than the minimum allowed peripheral thickness;
- the clear aperture which ought to be smaller than the maximum allowed clear aperture;
- the optical element glass, which ought to be inside the chosen glass triangle in the optical glass database.

If any of the boundary conditions are not fulfilled it is added to the merit function as "aberration". This so-called aberration is obtained as the difference between the boundary condition value and the minimum or the maximum allowed value. The calculation of aberrations, merit function and checking boundary conditions are repeated until no a new boundary condition violations is obtained i.e. the number of the boundary condition violations in the proceeding and the current procedure call is the same. After testing all boundary conditions the merit function is calculated and the DLS optimization can start.

The mathematical theory of the DLS optimization shows that the Jacobian matrix A – the matrix of the first partial derivatives of the ith aberration with respect to the jth constructional parameter and the matrix product $A^{T}A$ is needed. The calculation of the first partial derivatives is done with the finite differences method and it is explained in Section 8.3. After the calculation of the first partial derivatives and the matrix transformation, the solution of the linear equations system is needed and it is described in Section 8.4. During soluving the linear equations system the current value of the damping factor is used. The DLS convergence speed depends mostly on the correct determination of the damping factor so it is very important to select an appropriate value for the damping factor. The algorithm for the damping factor determination is given in Section 8.5.

With the selection of the correct value for the damping factor, the definition of a new optical system, having the smallest merit function value, which can be obtained in one iteration, is finished. However, a new optical system can have significant deviations from the boundary conditions. Therefore all boundary conditions ought to be checked to see if there are any violations. The process of boundary conditions checking is explained in Section 8.6. The DLS optimization is designed so that at the exit of each iteration it is necessary to obtain a new improved optical system, which has no new violations of the boundary conditions. The necessary conditions for the termination of the DLS optimization are given in Section 8.7.

8.2   Merit function

The merit function is the measure of the optical system performance. The merit function in the most general case can include the following:

− optical quality (for example the various kinds of aberrations as a measure of the optical system quality);

− physical realizability (for example every optical surface or optical element must be physically realizable which means that there must be no negative axial or edge thickness and that the radius of curvature for all optical surfaces must be greater than the clear aperture for these optical surfaces). The physical realizability is usually represented in the optimization algorithm through the boundary conditions;

− cost (for example the cost of material or fabrication).

A typical merit function includes terms to measure the optical system quality through the aberrations and physical realizability through the boundary conditions. The cost is usually controlled by limiting the types of materials used in optical system design, limiting the number of elements in the optical system and limiting the use of unusual surface types (like higher order aspherics). The cost is also determined by the manufacturing tolerances in the optical system. The author decided to use the compact merit function consisting of only optical system quality terms and boundary conditions violations. The merit function is defined as a sum of the squares of weighted operands. The operands included in the merit function are following:

− the paraxial values;
− the transverse aberrations;
− the longitudinal aberrations;
− the wavefront aberrations;
− the boundary condition violations.

The equation for the merit function definition is following:

$$\psi = \sum_{i=1}^{m} [\omega_i \cdot (f_i - f_{Ti})]^2 \tag{8.1}$$

where:

f_i is the calculated operands;

f_{Ti} is the desired or target operands. For the aberrations all target values are usually zero;

ω_i is the weighting factor for each operand.

The main characteristic of this merit function is the compact design, which uses small number of rays and extracts the maximum useful information from each traced ray. It is very well suited for the interactive optimization of optical systems. The best results with this merit function definition can be obtained with various kinds of objectives. The merit function consists of three groups of operands, which will be called aberrations for the sake of simplicity:

- the paraxial values;
- the aberrations;
- the boundary condition violations.

In the merit function there are nine paraxial values which are calculated as a finite difference between the calculated value and the desired (target) value. The paraxial values are following:

- the paraxial ray angle in the image space;
- the paraxial ray height at the last surface;
- the principal ray height at the aperture stop;
- the principal ray angle in the image space;
- the optical system thickness – the distance from the vertex of the first optical surface to the vertex of the last optical surface;
- the optical system length – the distance from the vertex of the first optical surface to the image plane;
- the distance from the object plane to the image plane;
- the glass thickness which is the sum of all lens thicknesses in the optical system;
- the defocus, which is a small displacement from the paraxial image position.

The number of aberrations in the merit function is variable and it is a function of the number of field angles and the wavelengths. It is possible to calculate aberrations with three field angles (on axis, $0.7 \cdot \omega_{max}$ and ω_{max}) or four field angles (on axis, $0.5 \cdot \omega_{max}$, $0.7 \cdot \omega_{max}$ and ω_{max}). It is also possible to calculate aberrations with one wavelength and the Conrady approximation for chromatic aberrations or with three wavelengths. The smallest possible number of rays is 17 and it is obtained for three field positions. This set of 17 rays is composed from one paraxial ray, two axial rays, two principal rays, eight oblique rays and four skew rays. When this set of 17 rays is traced through the optical system and only one wavelength is used, 38 aberrations are possible to be calculated. The following aberrations are calculated:

- for the axial and oblique rays the transverse ray aberrations and the Conrady chromatic aberration are calculated;
- for the principal rays the meridional and the sagital focal lines of the astigmatism and the distortion are calculated;
- for the skew rays the x and y components of transverse ray aberrations and the Conrady chromatic aberration are calculated.

When three wavelengths are used the number of calculated aberrations rises to 60. The following aberrations are calculated:

- for the axial and oblique rays the transverse ray aberration for each wavelength is calculated;
- for the principal rays the meridional and the sagital focal line of the astigmatism and the distortion are calculated;
- for the skew rays the x and y components of transverse ray aberration for each wavelength are calculated.

When four field positions are used the number of rays is 34. This set of 34 rays is composed from one paraxial ray, three axial rays, three principal rays, eighteen oblique rays and nine skew rays. When this set of 34 rays is traced through the optical system and only one wavelength is used, 78 aberrations are possible to be calculated. When three wavelengths are used, the number of calculated aberrations rises to 126.

It is very well known that finding an optical system with minimum aberrations usually depends on balancing, rather than removing, the aberrations from the optical system. In order to balance aberrations it is essential to make a sensible choice of weighting factors for the aberrations. The optical design program has the automatic procedure for assigning default values for the weighting factors which is based on the following criteria:

- in some cases it is better to have a sharp centre of the image, with the large flare, than a smaller flare with a less sharp centre. To attempt to achieve this the low aperture rays are given a larger weighting factor than marginal rays.
- it is permissible for the optical system performance at the maximum field angles to be worse than the performance at the smaller field angles. So it is advisable for the weighting factors at the larger field angles to be smaller than at the smaller field angles.
- the Conrady chromatic aberration is a wavefront aberration and it is much smaller than the transverse ray aberration. In order to correct both aberrations (the wavefront and the transverse ray aberrations) the wavefront aberrations ought to have much larger weighting factors.
- the marginal and sagital focal line of astigmatism are the longitudinal aberrations and normally will be larger than the transverse ray aberrations. So to correct all aberrations equally the longitudinal aberrations ought to have smaller weighting factors.
- it is very important not to attempt to control the aberrations that are uncontrollable. Examples of this are the distortion in eyepieces, the chromatic aberration in the Ramsden eyepieces and the astigmatism in the doublets. For these aberrations the weighting factors should have smaller values.

8.3   Differentiation

Calculating the first partial derivatives of the ith aberration with respect to the jth constructional parameter is possible to do in two different ways:

- the numerical differentiation based on the Newton's first interpolation formula or the Stirling's formula;
- the finite differences method.

During the implementation of the DLS optimization the author decided to use the finite differences method for all necessary calculation of the partial derivatives. The reasons for this decision are following:

- the sufficient precision is obtained with the finite differences method. It is possible to obtain greater precision with some numerical differentiation method but it does not bring any improvement in the optimization process as a whole.
- the finite differences method is much easier to implement in the computer program.

The finite differences method in calculation of the partial derivatives is also used by Kidger in [3].

During the calculation of the partial derivatives with the finite differences method each variable constructional parameter is changed for the small value and the aberrations are calculated. If a new calculated aberrations for the optical system with one changed variable constructional parameter are subtracted from the aberrations for the starting optical system with no variable constructional parameter changed then the partial derivatives of the aberrations are calculated with respect to the variable constructional parameter.

The quantity by which each variable constructional parameter is changed is the same for each parameter type, that is, there is a single increment for all radii, one for all axial distances, one for all refractive indices and one for all dispersions. When preparing input data for an optimization, care needs to be taken to ensure that sensible values are chosen for these increments, since if an increment is too small the DLS optimization will tend to make little change to any variable constructional parameter, whereas if an increment is too large the approximate linearity of the aberration changes with respect to the constructional parameter and the change will no longer be true. The program offers default values for all radii, axial distances, refractive indices and dispersions that are adequate for usual optical systems. The optical system designer should always start with those default values and see the optimization results. If the results are not satisfactory he should change default values slowly one by one.

8.4   Solution of the linear equation system

The methods for solving systems of linear equations are divided mainly into two groups:
- exact methods, which are finite algorithms for computing the roots of a system such as the Gaussian method;
- iterative methods, which permit obtaining the roots of a system to a given accuracy by means of convergent infinite processes.

The mathematical theory of methods for solving systems of linear equations is well known and can be found in any book about numerical mathematics such as [44] or [45].

The most common technique for solving a linear equation system is via an algorithm for the successive elimination of the unknowns. This algorithm is called the Gaussian method. The author has tested following methods that are all based on the Gaussian method:

– the classical Gaussian method;

– the Gaussian compact scheme;

– the Crout – Doolittle modification;

– the method of principal elements;

– the Khaletsky scheme.

After testing all these methods for solving linear equations the author decided to use the Crout – Doolittle modification of the Gaussian method. The reasons for this decision are following:

– easy programming of the method (the main part of the method consist of only three FOR loops);

– the economy in demand for the memory storage (the Crout – Doolittle modification does not record intermediate results);

– the sufficient precision of the method. It is theoretically proved that the Gaussian method and its variants become numerically unstable if the principal elements a_{ii} of the Jacobian matrix $\mathbf{A}$ are small or close to zero i.e. if the Jacobian matrix is singular or near singular. In the DLS optimization the Jacobian matrix cannot be singular because the damping factor is chosen so to prevent singularity of the Jacobian matrix.

8.5   Selection of the damping factor

The convergence speed of the DLS optimization depends on a suitable selection of the damping factor p. It is not easy to choose a satisfactory value for the damping factor when a new optical system is to be optimized, so the program attempts to optimize the selection of the damping factor. The basic flow chart for the damping factor selection is given in Fig. 8.2.

The selection of the damping factor is based on the dependence of the damping factor on the merit function ψ. Before starting the selection of the damping factor the program has calculated the necessary Jacobian matrix $\mathbf{A}$ and the matrix product $\mathbf{A}^T\mathbf{A}$. Also according to input data the program has decided to work with additive damping or multiplicative damping and changed accordingly the matrix product $\mathbf{A}^T\mathbf{A}$. For the additive damping the elements on the principal diagonal are added the constant term p^2 and for the multiplicative damping the elements on the principal diagonal are multiplied by the factor $\left(1 + p^2\right)$. After finishing all matrix calculations the program proceeds with the solving of the system of linear equations. For the first iteration, the program uses an initially arbitrary value of the damping factor p.

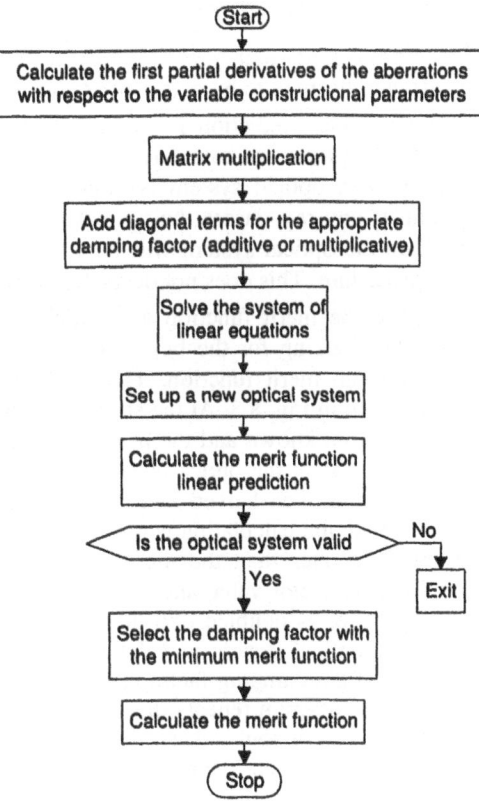

Figure 8.2. The basic flow chart for the damping factor selection

As a solution to the system of linear equations the set of variable constructional parameters is obtained, defining a new optical system. Then, the merit function linear prediction is calculated out of the initial aberrations, the partial derivatives and the constructional parameter changes using the following equation:

$$\Psi_{LIN} = \sum_{i=1}^{m} \left(f_{0i} + \sum_{j=1}^{n} a_{ij} \cdot x_j \right) \tag{8.1}$$

where:

Ψ_{LIN} is the merit function linear prediction;

f_{0i} is the initial value of the ith aberration;

a_{ij} is the first partial derivative of the ith aberration with respect to the jth constructional parameter;

x_j is the value of the constructional parameter change.

The merit function linear prediction will be correct if the constructional parameter changes are sufficiently small to be within the range of linearity.

After the merit function linear prediction is calculated the ray trace and the aberrations for a new optical system are calculated and a new optical system is tested to prove its validity. The optical system is valid if all emitted rays pass through the optical system. If any ray misses the optical surface or undergoes the total internal reflection, then the optical system will be considered invalid and the program returns from the procedure. This does not occur frequently.

When the aberrations and the merit function are calculated for a new optical system, the program can start looking for the best possible value of the damping factor which gives the minimum merit function. The detailed flow chart for the damping factor selection is given in Fig. 8.3. At the start of the optimization the best value for the damping factor is not known and the program uses the arbitrary initial value. For this damping factor value, the optical system and its merit function are calculated. The damping factor is then halved and the process of creating a new optical system and the merit function calculation is repeated. If the merit function of the second optical system is greater than the merit function of the first optical system, then the second damping factor value and the associated optical system are rejected. The first damping factor is doubled and the whole procedure of solving equations and creating a new optical system and the merit function calculation is repeated. When the direction of the damping factor change producing a decrease in the merit function is found, the program repeats this change of the damping factor (i.e. either halving or doubling its previous value), resolves the equations, creates a new optical system and calculates its merit function. If this third merit function is smaller than the previous merit function then the first damping factor value and the associated optical system are discarded. This process is repeated, the program always retaining two successive damping factor values with the associated optical systems that cause successively decreasing merit function values whilst calculating the third damping factor. At one moment this third merit function becomes greater than the preceding merit function and this is the moment when three merit functions span the minimum of the damping factor. When the range for the optimal damping factor value is found then the program tries to find the damping factor value for which the merit function has the minimum value. The merit function real dependence on the damping factor is not known in advance so approximate numerical methods must be used.

The most common approximation for this type function is the polynomial function. The author accepted the Kidger's proposal for the polynomial function [3]:

$$\psi = \psi_{LIN} + \delta\psi$$

$$\psi_{LIN} = \frac{1}{a + \dfrac{b}{p} + \dfrac{c}{p^2}} \tag{8.2}$$

$$\delta\psi = \alpha + \frac{\beta}{p^2} + \frac{\gamma}{p^4}$$

where

ψ is the real calculated merit function for a new optical system;

ψ_{LIN} is the merit function linear prediction for a new optical system;

$\delta\psi$ is the difference between the real calculated merit function and the merit function linear prediction for a new optical system;

$a, b, c, \alpha, \beta, \gamma$ are the unknown coefficients determined so that the real merit function is close to the approximated merit function.

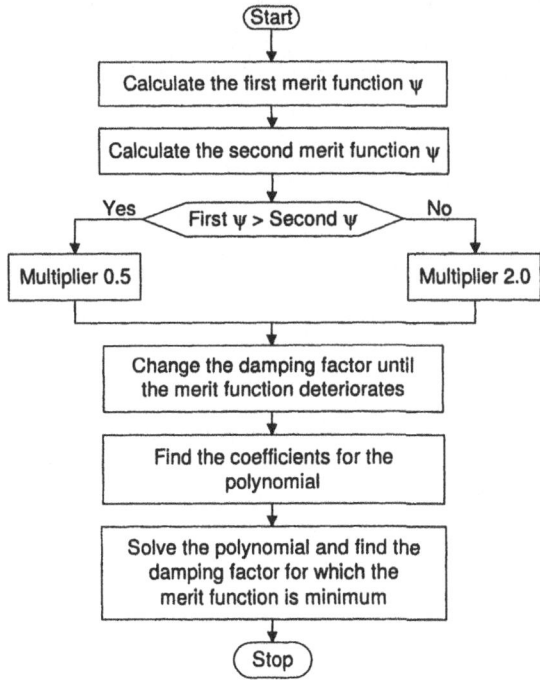

Figure 8.3. The detailed flow chart for the damping factor selection

The unknown coefficients $a, b, c, \alpha, \beta, \gamma$ are calculated if in Eq. (8.2) are included values of the real merit function and the merit function linear prediction, calculated for three values of the damping factor that span the minimum. The damping factor values will be marked $2p, p, \dfrac{p}{2}$ because the damping factor is calculated by either halving or doubling its previous value. The unknown coefficients $a, b, c, \alpha, \beta, \gamma$ are defined by the following equations:

$$a = \frac{1}{3 \cdot \psi_{LIN,1}} - \frac{2}{\psi_{LIN}} + \frac{8}{3 \cdot \psi_{LIN,2}}$$

$$b = p \cdot \left(\frac{5}{\psi_{LIN}} - \frac{1}{\psi_{LIN,1}} - \frac{4}{\psi_{LIN,2}} \right)$$

$$c = \frac{4 \cdot p^2}{3} \cdot \left(\frac{1}{2 \cdot \psi_{LIN,1}} - \frac{3}{2 \cdot \psi_{LIN}} + \frac{1}{\psi_{LIN,2}} \right) \qquad (8.3)$$

$$\alpha = \frac{1}{45} \cdot \left[-20 \cdot (\psi - \psi_{LIN}) + \psi_1 - \psi_{LIN,1} + 64 \cdot (\psi_2 - \psi_{LIN,2}) \right]$$

$$\beta = \frac{p^2}{9} \cdot \left[17 \cdot (\psi - \psi_{LIN}) - (\psi_1 - \psi_{LIN,1}) - 16 \cdot (\psi_2 - \psi_{LIN,2}) \right]$$

$$\gamma = \frac{4p^2}{45} \cdot \left[-5 \cdot (\psi - \psi_{LIN}) + \psi_1 - \psi_{LIN,1} + 4 \cdot (\psi_2 - \psi_{LIN,2}) \right]$$

If the variable q is defined as the inverse of the damping factor value ($q = \frac{1}{p}$) then Eq. (8.2) can be written in the form:

$$\psi = \frac{1}{a + b \cdot q + c \cdot q^2} + \alpha + \beta \cdot q^2 + \gamma \cdot q^4 \qquad (8.4)$$

In order to find the minimum value of the damping factor p, the first partial derivative is found and made equal to zero.

$$\frac{\partial \psi}{\partial q} = \frac{-(b + 2 \cdot c \cdot q)}{(a + b \cdot q + c \cdot q^2)^2} + 2 \cdot \beta \cdot q + 4 \cdot \gamma \cdot q^3 = 0 \qquad (8.5)$$

Eq. (8.5) is solved by the Newton – Raphson method and the obtained solution is the inverse of the damping factor value. This new minimum value for the damping factor is used for the optical system definition having the minimum merit function in this iteration.

8.6 Boundary conditions control

The boundary conditions may be defined as the constraints that are necessary to include into the optimization algorithm so that the optimization should produce a physically realizable optical system that satisfies all designer's requirements. Besides the optimization algorithm itself, the control of boundary conditions is considered to be one of the most important features of an optical system optimization program. In the author's implementation of the DLS optimization applied to the optical system optimization, there are two different methods of maintaining control of the boundary conditions: freezing to the constant minimum value or treating the boundary condition as an aberration and adding it to the merit function. There are also four different types of variables on which boundary conditions can be imposed: the axial separations between the optical surfaces, the optical surfaces radii of curvature, the optical system tolerances and the glasses.

8.6.1 Axial separations boundary condition control

The first group of boundary conditions are axial separations between the optical surfaces which can be either the axial thicknesses of the optical elements (lenses, prisms) or their distances from each other in the air and, for the real optical system, these must all have values greater than zero. The optimization algorithm, which works up on a mathematical model, may however find a theoretically improved system where some of these separations are negative or where the thickness of an optical element although greater than zero, is far too small for it to be manufacturable. Therefore the optical system designer must specify a set of tolerances, which will be imposed on the variable separations. The optical system designer can choose between a single tolerance applied to all axial thicknesses of the optical elements and another tolerance applied to all air separations. There is another approach to define separate values for all tolerances to be applied to each axial separation of the optical system.

When the optimization algorithm, at the end of iteration, finds a new improved optical system, it tests new values found for all variable axial separations against their tolerance values. If any of variable axial separations is smaller than the predefined tolerance values a new optical system is considered to be unsatisfactory and it is rejected. The variable axial separations whose values have become too small are 'frozen' i.e. they are not allowed to vary. The optimization algorithm returns to the optical system from the start of the current iteration and in this optical system all values of the newly frozen axial separations are replaced by their predefined tolerance values. For the adjusted optical system the ray trace is calculated in order to find a new set of aberrations. The Jacobian matrix is reduced in size by removing the first partial derivatives of the ith aberration with respect to the jth which has been frozen. A new reduced Jacobian matrix is calculated and the system of linear equations is solved. The next step is the calculation of a new minimum value for the damping factor and the optical system definition.

A new optical system is only acceptable if it has no axial separations violating the given tolerance values. If a new optical system does contain further axial separations the values of which are smaller than the predefined tolerance values, then the whole above-described process of freezing variables and calculating a new optical system is repeated until an acceptable system is produced.

The axial separations frozen in this way remain invariant and fixed at their predefined tolerance values during the remainder of current iteration and in all subsequent iterations. After certain number of iterations the optimization algorithm resets to the initial conditions meaning that all variables (axial separations) which were free at the start of the first iteration, but were subsequently frozen, are again allowed to vary. This 'defrosting' of variables is called 'memory clearance' and occurs during the optimization algorithm execution at an arbitrary point chosen by the optical system designer. The default value for the memory clearance is every five iterations.

8.6.2   Radii of curvature boundary condition control

The boundary conditions in the second group are related to variable radii of curvature. The most important manufacturing tolerances required are a reasonable thickness at the edge of each convex component to allow for mounting and maintenance of an acceptable diameter on all optical elements. These are controlled by imposing a tolerance on edge thickness of the positive (convex) optical elements. The quantity measured as edge thickness is the minimum optical path distance through any optical component whose edge is thinner than its axis. The optical system designer can choose between the single value for the minimum edge thickness tolerance applied to all convex optical elements or the separate tolerance value for each edge thickness of the convex optical elements.

During the ray trace through the starting optical system of each iteration, the shortest optical path distance for each space, the ray causing this shortest optical path distance and the wavelength in which it is traced are recorded. At the end of each iteration the minimum optical path distances are compared with the predefined tolerance values for each space where the edge is thinner than the axis. If any optical path distance is smaller than the required tolerance value it is considered as a violation of the boundary conditions and the difference from the optical path distance to the predefined tolerance value is calculated and added to the merit function as aberration. A new optical system that has violations of the boundary conditions is rejected and the optimization algorithm returns to the optical system at the beginning of the iteration and calculates a new merit function value which includes these additional violated condition aberrations. Next, the Jacobian matrix is expanded by adding the first partial derivatives of the violated condition aberrations with respect to the constructional parameters to each row of derivatives. Only rays causing these violated condition aberrations need to be traced at this stage. After that, the system of linear equations is solved and a new minimum value for the damping factor is calculated. Finally, a new and improved optical system is defined. Since the new improved optical system includes controlled edges in its solution, the edge thickness values which had previously been found to be less than their given tolerance values are now much nearer to the approved limits. This process called 'looping back' is repeated during the iteration until a new improved optical system without new violations is found.

The iteration is not considered to be complete until a new improved optical system is produced, containing no more new violations and with the merit function less than the merit function of the optical system at the start of iteration. This means that the optical system resulting at the end of each iteration is always an improvement over the starting optical system.

At the end of iteration the optimization algorithm decides whether to continue to control the same violations it has controlled during the iteration. It is very important to notice that the optimization algorithm controls only the violated condition aberrations such as edge thicknesses and system tolerances representing the real violations i.e. when their values are outside the given tolerance values. In order to stop further controlling the violated boundary condition two conditions must be fulfilled:

- the boundary condition value must be within the given tolerance, for example the edge thickness must be greater than the minimum allowed edge thickness;
- the forecast value of the boundary condition violation also must be within the given tolerance. This is important to prevent the reappearance of the boundary condition violation once when it is stopped beeing controlled.

The forecast value is calculated by assuming that the changes made to the starting optical system to convert it to the improved optical system (the solutions of the matrix equations) have linear relationship with the aberrations. The linear forecast of an aberration is given by:

$$f_f = f_s + \sum_{j=1}^{N} \frac{\partial f_s}{\partial x_j} \cdot x_j \qquad (8.6)$$

where:

f_f is the linear forecast of an aberration;

f_s is the same aberration with respect to the starting system;

$\dfrac{\partial f_s}{\partial x_j}$ is the change in this aberration produced by a change in the jth parameter only in the starting system;

x_j is the value of the change to the jth parameter found as a result of solving the matrix equations.

If both the calculated value of controlled violation with respect to a new improved optical system at the end of the iteration and its linear forecast computed by the given method lie within the given tolerance, then the violation is said to no longer exist and it is not controlled during the rest of iteration.

8.6.3 System tolerances boundary conditions

The third group of boundary conditions consists of paraxial conditions such as:

- the equivalent focal length;
- the magnification;
- the distance from the object plane to the image plane;
- the optical system thickness – the distance from the vertex of the first optical surface to the vertex of the last optical surface;
- the back focal distance – the distance from the last optical surface to the image plane;
- the optical system length – the distance from the vertex of the first optical surface to the image plane;
  These three last quantities may be constrained to achieve target values, or to be less than a predefined maximum or greater than a predefined minimum.
- the paraxial ray angle in the image space;
- the paraxial ray height at the last surface;
- the principal ray height at the aperture stop;
- the principal ray angle in the image space;
- the glass thickness, which is the sum of all axial thicknesses of all refracting optical elements and is used as a control over the amount of glass in an optical system. This is a simple means of preventing the optimization algorithm from making some optical elements unacceptably thick. Another way of dealing with the problem is to set the axial or edge thickness limits for those optical elements which have a tendency to become too thick as the maximum values instead of minimum values;
- the defocus which is small displacement from the paraxial image position;
- the maximum or minimum aperture at any optical surface.

Any or all of these boundary conditions can be controlled during the optimization. Violations of these boundary conditions are always treated in exactly the same way as the edge thickness violations.

8.6.4   Glasses

The optical system in general can have three groups of variable parameters: the first is the axial separations between the optical surfaces, the second is the radii of curvature and the third is the glasses of the optical elements. The radii of curvature, the separations and the aspheric coefficients are all naturally continuous variables, whereas the glasses represented by the refractive indices and the dispersions are not, so a suitable method of dealing with them has to be devised. The glasses (the refracting materials of an optical elements) are specified by their refractive indices in some mean wavelength and their dispersions for a chosen range of wavelengths. The refractive index n and the dispersion δn are treated as two independent variables, each assumed to be continuously variable in the ($n, \delta n$) plane within a given triangular region of which is determined by defining the refractive indices and dispersions of three vertices of the triangle. For the visible light the mean wavelength is usually the Fraunhofer's d spectral line representing the yellow helium light ($\lambda_d = 587.56$ nm) with dispersions in the range from the Fraunhofer's *C* spectral line representing the red hydrogen light ($\lambda_C = 656.27$ nm) to the Fraunhofer's *F* spectral line representing the blue hydrogen light ($\lambda_F = 486.13$ nm). Of course all wavelengths can be interactively changed before or after the optimization. The default values for the standard glass triangle are the well-known Schott glasses LakN9, SF4 and BK7. The optical system designer can interactively change one or all glasses from the standard glass triangle.

Within the glass triangle in the ($n, \delta n$) plane glasses are allowed to vary freely. If a new glass moves out of the defined glass triangle it is considered to be violating boundary conditions and it is controlled by treating the perpendicular distance from a new glass to the nearest boundary line of the glass triangle as a violated condition aberration that has to be constrained to zero. The refractive index and dispersion of the glass are still treated as variables.

The glasses in an optical system are controlled by computing the chromatic aberration of a ray traced through the optical system containing *K* glasses using the Conrady approximation formula in the form:

$$\delta\omega = \frac{1}{\lambda} \cdot \sum_{i=1}^{K} \delta n_i \cdot \left(\overline{D}_i - D_i \right) \tag{8.7}$$

where:

$\overline{D}_i$ is the optical path distance of the reference ray in the ith glass;

D_i is the corresponding optical path distance of the associated ray chromatic aberration $\delta\omega$ of which is to be found;

δn_i is the dispersion for the ith glass;

λ is the mean wavelength in the same measurement units as the optical system data (λ acts as a scaling factor to give results of the calculation in terms of wavelength).

The Conrady approximation formula for chromatic aberration involves, unacceptable errors in some optical systems. In its derivation the assumption is made that the paths of a ray traced through the optical system, in different wavelengths are close together; when it is not so, the Conrady formula is inapplicable. An alternative method of examining and controlling chromatic aberrations is to repeat the ray tracing through the optical system in a range of different wavelengths and to consider the same type of aberration in each of the chosen wavelengths. Repeating the ray tracing of an optical system in three different wavelengths increases the time taken by the ray-trace-dependent part of the program by a factor of three, which must be taken into account by the optical system designer in choosing between the essentially more accurate method and the Conrady formula.

8.7   Termination of the damped least squares optimization

When the optimization algorithm has created a new optical system having the smallest possible merit function for the given iteration it is necessary to check whether the optimization is continued or not. There are several criterions which can terminate the optimization process:

- when the merit function difference for two consecutive iterations is not greater than a particular small value. The default value is 0.01 and it can be interactively changed before or after optimization.

- when the difference between the real calculated merit function for a new optical system and the merit function linear prediction for this optical system is smaller than a particular small value. The default value is 0.01 and it can be interactively changed before or after optimization.

- when the number of executed iterations is greater than the maximum number of iterations. The default value for the maximum number of iterations is 1000 iterations and it can be interactively changed before or after optimization.

It is important to notice that it is necessary to fulfil only one criterion to terminate the optimization process.

Chapter 9

Adaptive steady-state genetic algorithm implementation

9.1 General description of the adaptive steady-state genetic algorithm

The adaptive steady-state genetic algorithm (ASSGA) represents one of possible genetic algorithms for the optical system optimization. The genetic algorithms are a general search technique and they are not primary designed for the technical system optimization. On the other hand the ASSGA is specially designed for the technical system optimization. It has some differences from the rest of genetic algorithms, which makes the ASSGA a good algorithm for optimization. The theory of the ASSGA along with other genetic algorithms is given in Chapter 3.

The first big difference between the ASSGA and the classical genetic algorithms is in the representation of individuals in the population. The classical genetic algorithms use the bit strings and the ASSGA the real numbers. In technical system optimization it is very important to use representation of the system that is developed and tested and not to develop a new representation that has no connection with the technical system. In the case of the optical system optimization every optical system is defined with the record that contains the basic constructional parameters like:

- radii of the curvature for all optical surfaces;
- distances between the optical surfaces;
- glasses from which the optical elements (lenses, prisms) in the optical system are made. The glasses are represented by the glass name from the appropriate glass manufacturer, the index of the refraction for three spectral lines and the dispersion. The index of refraction and the dispersion are used for the optical system calculation.
- clear radii for each optical surfaces;
- aspherical surface coefficients for each aspherical optical surface if there are aspherical surfaces in the optical system exists;

In order to define an optical system completely it is necessary to, behind the basic record with the construction data, define a record of the ray data i.e. the coordinates of the selected set of rays on the entrance reference plane. This set of rays simulates the passage of the light through the optical system and it is used for calculating all optical system properties like the aberrations, the optical path difference, the spot diagram, the modulation transfer function and the merit function.

The three records are used to completely define an optical system as an individual for the ASSGA optimization:

- the first record with the construction data;
- the second record with the ray data;
- the third record with the aberrations and the corresponding waiting factors that define the merit function.

Each optical system to be accepted to the population must fulfil all boundary conditions like:

- an optical system must have a strictly defined effective focal length;
- it must be possible to calculate the coordinates of the predefined set of rays for the optical system;
- all radii of curvature and all distances must be physically realizable;
- all minimum lens thickness at the axis and at the periphery must be greater than predefined values;

When a new-formed optical system satisfies all boundary conditions then it can be accepted in the population. This means that the population consists only of optical systems that fulfil all boundary conditions and that can be physically realized in every moment and that represent valid candidates for the optimization problem solution.

The population size is a very important parameter in order to find an optimum solution by the genetic algorithm. If the population is too small then there are not enough different individuals, which will make possible for the genetic algorithm to search the optimization space successfully, escaping the local minima and looking for the global minimum. Too big population is not good as well because the larger number of individuals than necessary extend the time needed for the optimization to find the optimum solution. The time needed for creating and evaluating each individual is not negligible so the number of individuals in the population should be kept to a reasonable size. There is no theoretically defined way for determining the optimal population size. The population size is mostly defined experimentally or by the designer's experience with the optimization method. The population size is the function of the optical systems complexity. If the optical system is simple one like the doublet or the triplet, the population size can be smaller e.g. such as 100 to 300 individuals and if the optical system is more complex like the double Gauss objective, the population size must be larger for at least 500 individuals. Because of its great importance in the optimization the population size is one of the most important parameters, which the designer of optical systems must specify for each optimization run.

The second big difference between the classical genetic algorithm and the ASSGA applied to the optical system optimization is the method of generating the initial population. In the classical genetic algorithm generating the initial population is completely random while in the optical system optimization completely random initialization must be adjusted to the specific way of the optical system designing and optimization. When optimizing an optical system there is always a starting point i.e. the optical system designer always haa an optical system that he tries to improve during the optimization. The designer always starts from the optical system that can represent a better (the optical system closer to the optimum) or a worse (the optical system further from the optimum) starting point. In this initial optical system only certain set of construction parameters are allowed to be changed during the optimization. It is never allowed to change all the construction parameters. It is usual to allow changing the following construction parameters during the optimization:

— some radii of curvature;
— some separations between the optical surfaces;
— some glasses from which the optical elements are made;
— some aspherical coefficients of optical surfaces if there are aspherical surfaces in the optical system.

The reason why to insist on some construction parameters and not on all construction parameters is because in the optimization it is not allowed to change the type of the optical system. This means that the arrangement of optical components must be constant i.e. if the doublet is optimized it will be also the doublet at the end of optimization but it will have some of the radii of curvature changed, as well as the separations between the optical surfaces and the glasses which will allow it to have smaller aberrations and the merit function. It is important to notice that the doublet will never transform to e.g. the triplet during the optimization nor the converging and diverging lenses will change their places. In order to satisfy all this conditions the process of generating optical systems is quite complicated and many conditions must be fulfilled. A detailed description of generating optical systems for the initial population is given in Section 9.2.

The third big difference between the classical genetic algorithm and the ASSGA is that the classical genetic algorithm uses the generational model and the ASSGA the steady-state model without duplicates. The generational model means that the new-formed generation replaces completely the existing generation. The steady-state model without duplicates means that the population consists from different individuals. In this population only one individual is changed. This changed individual must pass the series of tests to become a valid optical system, must be different from the rest of optical systems in the population and ought to have the merit function smaller than the biggest merit function of the optical system from the population to be accepted to the population. Then if all these conditions are fulfilled, the changed optical system replaces the worst optical system with the biggest merit function in the population. The optical system with the smaller merit function has smaller aberrations and it is a better optical system. The steady-state model enables greater difference in optical systems in the population and that means the greater speed in the search of the optimization space and the quicker convergence to the optimum solution. With the generational model, in the population there are usually duplicates of the very good individuals, which reduces the diversity in the population and the speed in the search of the optimization space. The second good property of the steady-state model is that none of good individuals, that can represent possible solution to the optimization problem i.e. optimum optical system, is lost because only the worst individual in the whole population is changed. At the classical generational model there is a theoretical chance that some very good individuals can be lost during the transformation from one generation to another generation.

The fourth big difference between the classical genetic algorithm and the ASSGA is in a number of different genetic operators. The classical genetic algorithm usually has only two genetic operators: one point crossover and the mutation. The ASSGA has five genetic operators that are completely independent from each other and every one is executed with its own probability. Only one genetic operator can be executed on a single individual. The following genetic operators are used in the ASSGA:

– the uniform crossover is the generalisation of the single point crossover and with the uniform crossover all imperfections of the single point crossover are corrected. In the case of uniform crossover the process of crossing over takes place for each position in the individual. It is randomly decided which parent will contribute to the value of offspring. The default value for the uniform crossover probability is 20%. This means that in only approximately 20% reproductions the genetic operator will be uniform crossover. This is only the initial value for the uniform crossover probability, which can be modified during the optimization with the genetic operator adaptation. The optical system designer can specify interactively before or after the optimization desired initial value for the uniform crossover probability. This and all other optimization parameters are held in a separate data file.

– the average crossover is the crossover that takes two parents and forms only one offspring that is the result of averaging the corresponding positions of two parents. The default value for the average crossover probability is 20%. This is only the initial value for the average crossover probability which can be modified during the optimization with the genetic operator adaptation. The optical system designer can specify interactively before or after the optimization desired initial value for the average crossover probability.

– the floating point number mutation is the mutation that randomly changes one of the variable construction parameters. The default value for the change of the selected variable construction parameter is within the limits of 10 times the original value of the construction parameter. Since the mutation is a potentially destructive genetic operator there are two probabilities that must be fulfilled for the mutation execution. The first probability is the probability that the mutation is selected as a genetic operator for reproduction and the default value is 20%. This is only the initial value for the probability, which can be modified during the optimization with the genetic operator adaptation. The second probability is the probability of the mutation execution once when the mutation is selected as a genetic operator and the default value is 50%. It is important to notice that the change of the selected variable construction parameter, the probability for selecting mutation as a genetic operator and the probability of the mutation execution can be specified interactively before or after the optimization.

– the big number creep is the mutation that randomly changes one of the variable construction parameters. The big number creep is very similar to the floating point number mutation with the only one difference. The difference is in the limits of the construction parameter change. For the floating point number mutation the default value for the limit is 10 times the original value of the selected variable construction parameter and for the big number creep the default value is 4 times the original value of the selected variable construction parameter. The big number creep also has two probabilities. The first one is the probability of selecting the big number creep as a genetic operator for reproduction and the default value is 20%. This is only the initial value for the probability which can be modified during the optimization with the genetic operator adaptation. The second probability is the probability of the mutation execution once when it is selected as a genetic operator and the default value is 70%. As in the floating point number mutation all parameters for the big number creep can be specified interactively before or after the optimization and they are placed with other optimization parameters in a separate data file.

– the small number creep is the mutation that randomly changes one of the variable construction parameters. The small number creep is very similar to the big number creep. The parameters for the small number creep are the following: the default value for the change of the selected variable construction parameter is in the limits of 2 times the original value of the construction parameter; the default value for the probability of selecting the small number creep as a genetic operator for reproduction is 20% and the default value for the probability of executing the mutation once when it is selected as a genetic operator is 70%. The default value for the probability of selecting the small number creep as a genetic operator for reproduction is only an initial value which can be modified during the optimization with the genetic operator adaptation. All these parameters for the small number creep can be specified interactively before or after the optimization and they are placed with other optimization parameters in separate data file.

It is important to notice once more that all these probabilities can be adapted to suitable values during the optimization. The process of adapting values for the probabilities reflects the current state of the optimization i.e. the current shape of optimization space. All starting values for the probabilities, although they are variable during the optimization are very important and they are input parameters that the optical system designer must define for each optimization run. The author, during his research with various types of objectives, found that the above default values for the optimization parameters are most appropriate values for the optimization parameters.

The basic flow chart diagram for the ASSGA is shown in Fig.9.1. From Fig. 9.1 it can be seen that this is a classical flow chart diagram of the genetic algorithm. The first step is generation of the initial population. This is a rather complex operation because the optical system optimization has a lot of specific things that must be fulfilled. The whole process of generating initial population is described in Section 9.2. When the whole population is initialized the genetic optimization is started with several steps that are repeated in the loop. The first step is the parent selection, which is the process of selecting two parents for creating a new offspring, and this is described in Section 9.3. After selecting the parents, one of genetic operators which will be applied to the parents in order to form one or two offspring, must be selected. When a new offspring is formed, i.e. a new optical system is created, it must be tested to find if it satisfies all boundary conditions. A new optical system can be accepted to the population only if it satisfies all boundary conditions and has the merit function that is smaller than the biggest merit function of the worst individual in the population. This part of selecting genetic operator and creating and testing a new optical system is described in Section 9.4. The final part of genetic optimization is the genetic operator adaptation, a process of adapting probabilities of genetic operators to be selected according to their performance. The genetic operator that creates good individuals, i.e. the optical systems with the small aberrations and the small merit function, get the larger chance to be selected. The whole process of genetic operator adaptation is described in Section 9.5.

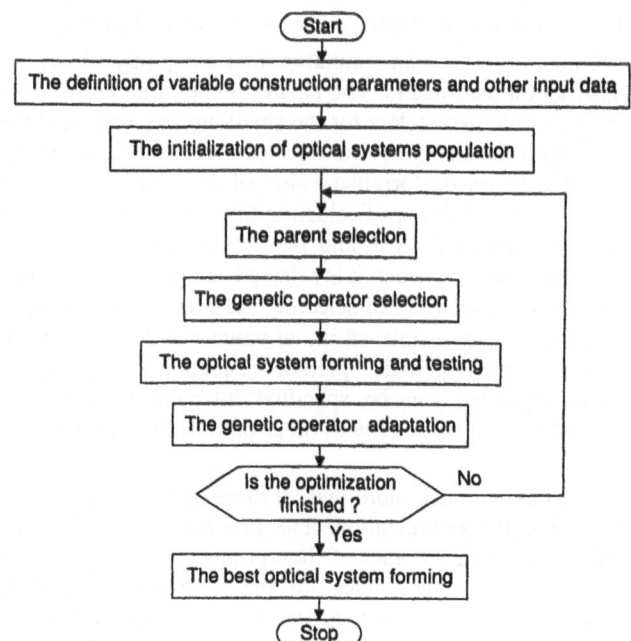

Figure 9.1. The basic flow chart diagram for the ASGGA optimization

At the end of each generation calculation all criterions for the termination of genetic optimization are checked for fulfilment. The criterions for the termination of genetic optimization are the following:

– the number of generations i.e. the maximum number generations which will be calculated by the genetic algorithm before the optimization is finished;

– the population convergence which means that the optimization algorithm checks whether thr whole population is converged to the single value i.e. to the single optical system. If the difference for the calculated optical system merit function for the best and the worst individual in the population is less than 5% of the merit function value then it is considered that the population is converged to the solution and there is no point in further calculating new generations.

9.2 Generating the initial population

Generating the initial population is the first step that must be executed in every genetic algorithm. The optical system optimization is rather specific and the optical systems cannot be completely randomly generated for the initial population. The strict procedure must be followed which will enable the generation of the optical systems that have some predefined characteristics. The flow chart diagram for initial population generating is shown in Fig.9.2.

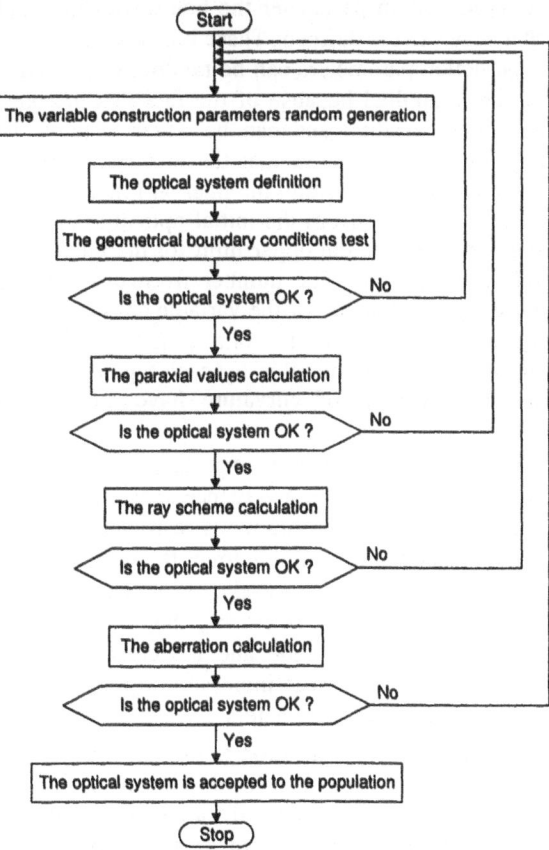

Figure 9.2. The flow chart diagram for initial population generating in the ASSGA optimization

All optical systems for the initial population are generated from the starting optical system. For every optical system optimization the initial optical system must be known i.e. selected by the designer of optical systems. The initial optical system must be a valid optical system meaning that the optical system must satisfy all boundary conditions. For this initial optical system some of construction parameters are made variable and some remain fixed. The designer of optical systems will make decision which of construction parameters will be variable and which will be fixed according to his experience and intuition.

The new optical system for the initial population is generated in the following way:

– the fixed construction parameters from the initial optical system are copied;
– the variable construction parameters are randomly generated beginning from the initial optical system construction parameters;

For each variable construction parameter the boundaries in which the variable construction parameter may exist is defined. These boundaries represent the interval in which variable construction parameters can be randomly selected. The following method for the selection is applied because of the fact that the positive radius of curvature must always stay positive and can be never negative because it will destroy the optical system conception, or that the lens thickness must be always positive in order to be physically realizable: the interval limits are calculated so that the variable construction parameter from the initial optical system is multiplied by the predefined constant and become the upper limit for the interval while the lower limit for the interval is set at zero. The number from this interval is generated randomly and it represents a new value for the variable construction parameter. The optical system designer can select to use one constant value for all variable construction parameters or one constant value for each type of variable construction parameters meaning that all radii of curvature have one constant value, all separations the other value and so on.

The predefined constant value is a very important factor in the possibility of finding the global minimum and in the convergence speed. If the constant value is too small then the number of possible solutions is restricted. It is possible that the optical system that represents the global minimum or the optical system that leads to the global minimum lies outside the allowed area of the variable construction parameter change. If the constant value is too big then the range for the variable construction parameter change is too wide and the creating of the new optical systems becomes completely random. It is impossible to define the exact constant value because it is a function of many parameters.

The best way for constant value definition is the experience with the ASSGA optimization and the experiments with the selected optical system. This means that several values for the constants are selected and for each selected value quite a few ASSGA optimizations are executed. Because of great importance that the constant value has in the ASSGA optimization the optical system designer must define it or accept the default value for the every ASSGA optimization run. The best approach is to accept the default value and to see the optimization results. Because of its random nature several optimizations must be executed before the optical system designer can judge anything. If the constant value is inadequate then the designer can change it.

After the variable construction parameter random selection the parameter must be checked. If the variable construction parameter is:

- the radius of curvature then the following criterions must be fulfilled:
    - the radius of curvature must be different from zero;
    - the absolute value for the radius of curvature must be greater than the clear aperture.
- the separation between the optical surfaces then the following criterions must be fulfilled:
    - the separation between the optical surfaces must be different from zero;
    - the separation between the optical surfaces must be greater than the minimum allowed axial separation between the optical surfaces;
- the glasses of optical elements are chosen randomly from the optical glass database from the one of several world famous manufacturers of the optical glass. In glass selection there are no conditions because all glasses in the database are good and can be selected randomly;
- if aspherical surfaces exist in the optical system and if they can be changed then all coefficients that describe the aspherical surface must be different from zero.

The initialization and selection of the optical system variable construction parameters is done until all criterions are fulfilled. When a new optical system is created for the initial population the detailed check must be performed before the new optical system can be accepted to the population. Performing the detailed optical system checking is very important because the optimization algorithm can manipulate only with the valid optical systems that must fulfil numerous geometrical and optical conditions.

The geometrical condition checking is in the essence checking if the peripheral lens thickness is greater than the predefined minimum lens thickness. If any of the predefined lens thicknesses is smaller than the minimum lens thickness then the whole optical system is rejected and a new optical system is generated and tested. The process of testing and generating a new optical system is repeated until all population places are filled.

The optical condition checking consists of the following:

- the paraxial checking of the optical system;
- the ray scheme calculation;
- the aberration calculation.

For each optical system that is created the basic paraxial properties like the effective focal length and the back focal distance ought to be calculated. As the optical systems are created by random selection of the variable construction parameters there are large differences in the effective focal length between the optical systems. The optimization of the optical systems with the specified effective focal length is interesting: all the optical systems must be recalculated to have the desired effective focal length. This recalculation is done by changing the radius of curvature for the last refracting or reflecting surface of the optical system. When the optical system is recalculated to the desired effective focal length the calculation of the basic paraxial properties are performed once more. If during the calculation of the basic paraxial properties some errors occur this means that the optical system is not valid and it cannot become the member of the optical system population. It is rejected and a new optical system is generated and tested. The process of testing and generating new optical system is repeated until all population places are filled. Until now only the effective focal length was considered but also the back focal distance is very important and must be greater than zero. This is because some kind of detector is usually placed in the focal plane of the objective (the optical system with the finite focal length).

For the determination of the optical system quality (the aberration calculation and the merit function calculation) it is not enough to specify all the construction data for the optical system but it is necessary to define a set of rays that pass through the optical system. The coordinates of these rays on the reference plane are called ray scheme and they are used for ray tracing through the whole optical system. The process of calculating the ray scheme is the iterative process because the set of rays usually needs to be traced relatively to the principal ray which must pass through the centre of the aperture diaphragm. When the aperture diaphragm is in the optical system the position of the principal ray at the reference plane is not possible to be calculated directly (analytically) but it must be done iteratively by reducing the difference between the principal ray coordinate at the aperture diaphragm and the centre of the aperture diaphragm. All other ray coordinates at the reference plane are calculated relatively to the principal ray according to the predefined scheme.

If during the ray scheme calculation some errors occur this means that the optical system is not valid and it cannot become the member of the optical system population. It is rejected and a new optical system is generated and tested. The process of testing and generating a new optical system is repeated until all population places are filled.

The final optical condition checking is the aberration calculation. For each optical system from the population the aberrations that form the merit function are calculated. It is important to notice that the possibility or impossibility to calculate the aberrations is not essential for the optical system. If some aberration cannot be calculated the predefined large value is assigned and incorporated in the merit function. The problem may appear if an error occure, making further optical system calculation impossible: then the optical system is rejected and a new optical system is generated and tested. The process of testing and generating a new optical system is repeated until all population places are filled.

9.3   Parent selection

When the whole initial population is generated and tested the genetic optimization can start. The optimization is the iterative process and each iteration is called generation. In each generation the optimization algorithm must choose parents for creating new offspring, test this offspring and if all conditions are fulfilled accept this offspring. So the first step in each generation calculation is the parent selection. The flow chart diagram for the parent selection is shown in Fig. 9.3.

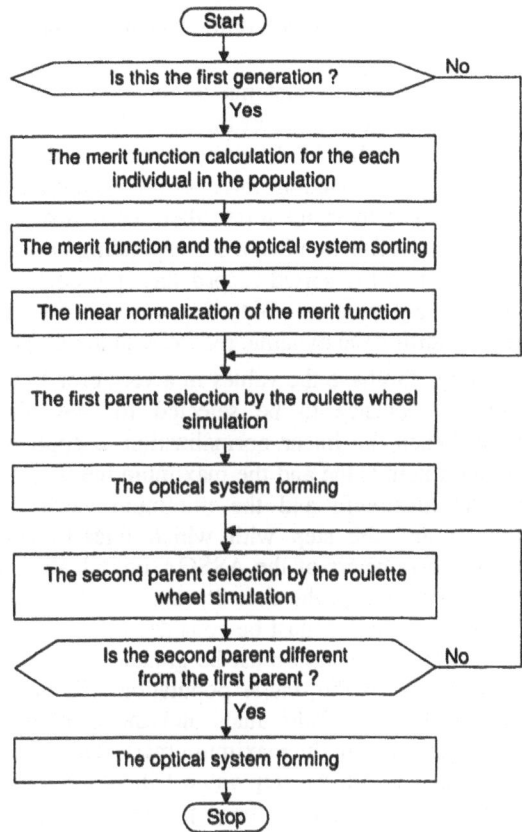

Figure 9.3. The flow chart diagram for the parent selection in the ASSGA optimization

The first step after the initial population generation and testing is calculation of the merit function for each optical system. The merit function is inherited from the classical damped least squares optimization and it is described in Chapter 8.

After the merit functions for all optical systems in the population are calculated then the optimization algorithm can proceed with the selection of two parents for the reproduction. The ASSGA optimization uses the roulette wheel selection, which is a standard selection mechanism for the genetic algorithms. The classical roulette wheel selection is designed for the maximization problems because the individual with the biggest merit function has the greatest chance in being selected for the parent. The optical system optimization is the classical minimization problem because the better optical system has the smaller aberrations and the smaller merit function. If the roulette wheel selection is applied without modification to the optical system optimization, the selection mechanism will give the greatest chances for the selection to the worst individual instead to the best individual. Therefore, it is obvious that some merit function mapping is necessary do that the best i.e. the optical system with the smallest merit functionn has the greatest chance in the selection for the parent.

The merit function for all optical systems in the population is stored in a dynamic record. The first thing to do is to sort this record with the ascending order so that the first merit function value in the sorted record is the minimum merit function value and the last merit function value is the maximum merit function value. When the merit function values are sorted then the optical systems must be sorted the same way. Each merit function has a corresponding optical system and if the merit function is the first in the sorted record then the corresponding optical system must be the first in the population. With this way it is possible to know the best optical system only by sorting the dynamic record with the merit functions.

The merit function values can have the values in a very broad range. In order to give each merit function a chance to be selected for the parent the linear normalization is applied. When the linear normalization is applied the minimum merit function gets the maximum value and the maximum merit function value gets the minimum value. The maximum and the minimum value for the linearly normalized merit function and the step with which linearly normalized merit function is reduced are the parameters of the ASSGA optimization and the optical system designer must select or accept the default value for each optimization run. When these parameters are specified it must be considered that the maximum value for the linearly normalized merit function must be greater than the population size because if not so, the mapping of the whole population will be impossible. For example: if the population has 200 individuals and the mapping i.e. the linear normalization ought to be ranked from the maximum merit function value 100 to the minimum merit function value 1 with the step value 1 then it is impossible because only 100 function places are available and, the population beeing 200 individuals, only one function place corresponds to each individual. The minimum value for the linearly normalized merit functions must be one or more because at the parent selection by the roulette wheel simulation the individuals having zero or negative merit functions are not allowed. This rule also applies to the step for linear normalization, which must be a positive integer greater than or equal to one.

The merit function values after linear normalization are placed in a separate dynamic record in the following way:

- if the population has 100 individuals and the linear normalization of the optical system merit functions is done from the maximum value 100 to the minimum value 1 with the step 1 then the individual with the minimum optical system merit function will have the maximum value of the linearly normalized optical system merit function which is 100;

- the next individual with the optical system merit function will have the linearly normalized optical system merit function, which is 99. In the dynamic record this individual is placed with a value which represent the sum of the linearly normalized merit function for this individual and the sum of all linearly normalized merit functions for the proceeding individuals. In this example this is $99 + 100 = 199$.

When the linearly normalized merit functions are calculated for the whole population then the parent selection by the roulette wheel simulation can be executed. The procedure for the roulette wheel parent selection is as follows:

Randomly select one number between one and the sum of linearly normalized merit functions for all individuals (optical systems) in the population. Each member of the special dynamic record with the linearly normalized merit functions is checked whether it is greater or not from the selected random number. Whenever the random number is selected checking is started with the first member in the special dynamic record. If the selected random number is greater than the linearly normalized merit function, checking is proceeded with the next linearly normalized merit function. If the selected random number is less than the linearly normalized merit function, the optical system with this merit function is selected for the parent.

Two parents, which are different optical systems, are always selected. This is done because it is not known in advance whether the selected genetic operator will demand one or two parents.

9.4   Creating and testing a new optical system

When the parents, which represent two different optical systems, are selected the optimization algorithm can proceed with the reproduction part of the algorithm. This is a very important part of the optimization algorithm because new optical systems are created and with the creation of new optical systems the optimization algorithm can search the optimization space toward the global minimum. The reproduction part of the optimization algorithm has several steps:

–   the genetic operator selection;

–   creating new optical system;

–   testing the optical system.

The detailed flow chart diagram for the reproduction part of the ASSGA optimization algorithm is shown in Fig. 9.4. The first step in the reproduction part of the ASSGA optimization algorithm is the genetic operator selection. As shown in Section 9.1 the ASSGA optimization has several genetic operators. They belong to the crossover operators and to the mutation operators.

The genetic operators are:

–   the uniform crossover;

–   the average crossover;

–   the floating point number mutation;

–   the big number creep;

–   the small number creep.

The first two genetic operators belong to the family of crossover operators and the last three genetic operators belong to the family of mutation operators. Each genetic operator has the probability for selection to be a genetic operator which will produce the new optical system. Since there are several genetic operators, the roulette wheel simulation it is best to be used for the genetic operator selection.

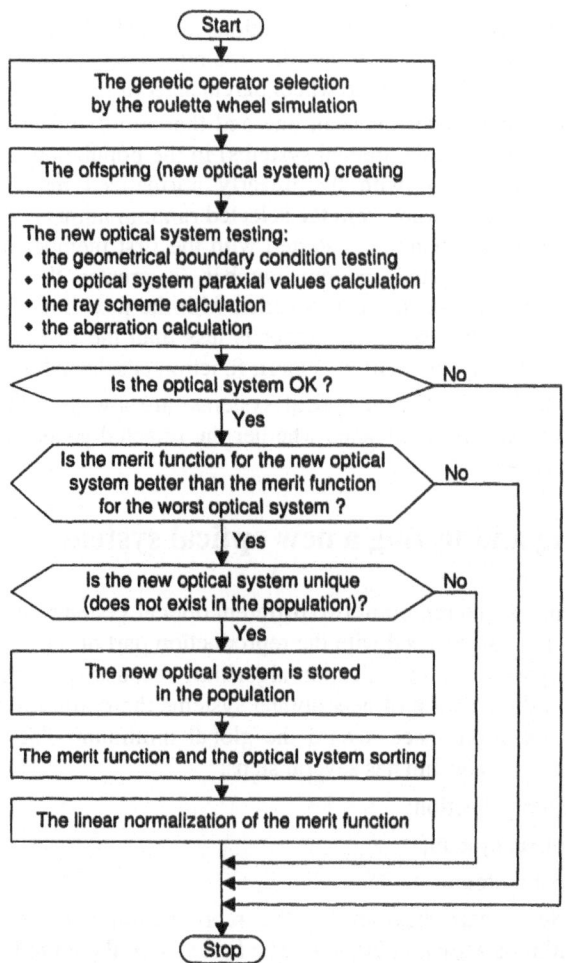

Figure 9.4. The flow chart diagram for the reproduction part of the ASSGA optimization

The roulette wheel simulation has one good property: the good genetic operators which produce the optical systems with the small merit function, have the greater chance to be selected as a genetic operator for reproduction. It ought to be noticed that the sum of probabilities for all genetic operators is always 100. The probabilities for all genetic operators are stored in the dynamic record in the following order:

– the probability for the first genetic operator is stored as it is;

– the probability for the second genetic operator is stored as the sum of probabilities of the first and the second genetic operator;

– the probability for the fifth genetic operator is stored as the sum of all probabilities for five genetic operators and it is always 100.

The roulette wheel simulation for the genetic operator selection is similar to the roulette wheel simulation for the parent selection. The procedure for the roulette wheel genetic operator selection is the following one:

Randomly select one number between 1 and 100. If this number is greater than the first probability in the dynamic record then the second probability is compared with the selected random number. The comparison is done until the probability greater than the selected random number, is found. The genetic operator that corresponds to this probability is selected to be a genetic operator for the reproduction.

After the genetic operator is selected it is applied to the selected parents. As the outcome of this reproduction the following results can be obtained:

− no offspring is created because all mutation genetic operators have two probabilities: one probability to be selected as a genetic operator for reproduction and other probability to be executed;
− one offspring is created with the average crossover and all mutation genetic operators if both probabilities are fulfilled;
− two offspring are created with the uniform crossover.

Each optical system created through the reproduction ought to be tested. The process of the optical system testing is a complex process and it is explained in details in Section 9.2. Only necessary steps will be stated here without any further description:

− the optical system paraxial values calculation like the effective focal length and the back focal distance;
− the optical system scaling to obtain the given effective focal length;
− the geometrical boundary condition testing for the optical system;
− the ray scheme calculation;
− the aberration calculation.

If the error in calculations occurs during testing any of these conditions, the optical system is rejected because it does not fulfil all necessary conditions. To be accepted in the population the optical system must fulfil the following conditions:

− the optical system has the smaller merit function than the worst optical system with the biggest merit function in the population;
− the optical system is unique i.e. there is no copies of the optical system.

When the optical system is accepted to the population, the population of optical systems and their corresponding merit functions must be sorted in the ascending order. Furthermore, the linear normalization of the optical system merit functions must be performed in order to enable the selection of new parents and a new genetic operator for the next reproduction.

9.5   Genetic operator adaptation

In most versions of genetic algorithms the genetic operators have fixed probabilities of execution. A large number of researches are conducted (some are described in Chapter 3) in order to find optimum values for the genetic operator probabilities. From the obtained results it is possible to conclude that the genetic operator probabilities are the function of many variables and they cannot be set in advance for any problem. It can be also noticed that the genetic operator probabilities ought to be variable during the optimization. With the variable genetic operator probabilities the most successful genetic operators are executed more frequently meaning that the genetic operators, which produce the optical system with the smallest aberrations and the merit function, are executed more frequently. In the ASSGA optimization the algorithm for the genetic operator adaptation is included. The detailed flow chart diagram for the genetic operator adaptation is shown in Fig. 9.5.

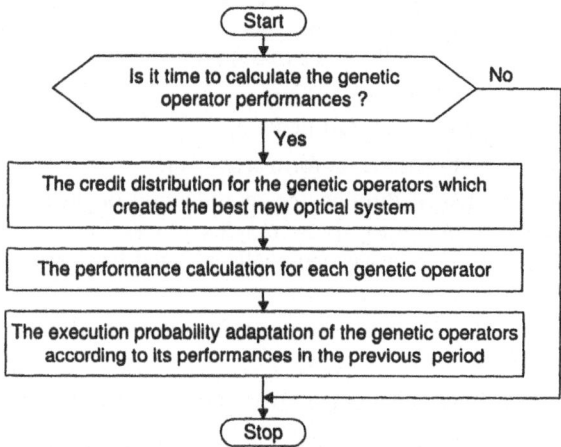

Figure 9.5. The flow chart diagram for the genetic operator adaptation in the ASSGA optimization

The detailed theory of the genetic operator adaptation is given in Chapter 3. The genetic operator adaptation is a complex process consisting of two steps:
– the genetic operator performance calculation;
– the genetic operator adaptation.

The performance of the genetic operators is calculated according to the optical system quality represented by the aberrations and the merit function they create. If the genetic operator creates better optical systems with the smaller merit function it will have a greater sum of credits and more chances to be executed and to obtain even more credits.

When a new optical system is created it is very important to write down the parents and the genetic operator that created new optical system in a separate dynamic record. This is important for the adequate credit calculation of genetic operators. When a created optical system is better than the best optical system in the population, credits should obtain not only the genetic operator that created this optical system but also the genetic operators that created parents and make it possible for the selected genetic operator to produce the best optical system. In the author's implementation of the ASSGA optimization the assigning credits to the genetic operators, which produced the best optical system in that moment go to the past three generations. The genetic operator that created the best optical system in that moment obtains the credit, which is equal to the difference between the merit functions for the new best optical system and the old best optical system. All other genetic operators that contributed to the creation of the best optical system in that moment obtain 15% of the credit that is assigned to the genetic operator which created the best optical system at that moment.

After a certain interval all credits obtained by each genetic operator are needed to be added together and divided by the number of optical systems created by each genetic operator. The interval size is a parameter of the optimization algorithm and can be changed interactively before or after the optimization run. The default value for the interval size is 50 generations. The performance of each genetic operator is calculated using this procedure.

The next step is the adaptation of the genetic operator probabilities. The detailed algorithm for the adaptation of the genetic operator probabilities is given in Chapter 3 and there is no specific data connected to the optical system optimization so it will not be described here. It is important to notice that the sum of all genetic operator probabilities must be equal to 100 before and after the adaptation of the genetic operator probabilities. When the adaptation of the genetic operator probabilities is calculated there is great chance that the sum of the genetic operator probabilities is not exactly 100. Then all the genetic operator probabilities must be adjusted to give exactly 100 as the sum of all genetic operator probabilities.

9.6   Adaptive steady-state algorithm variable parameters

The complete description of the adaptive steady-state genetic algorithm and its most important parts are given in the proceeding sections. The ASSGA is a quite complex optimization algorithm with many variable parameters thus enabling greater flexibility of the optimization algorithm. The optical system designer can specify interactively all variable parameters placed in the separate data file before or after the optimization. The following ASSGA optimization parameters can be modified:

- the population size is the number of individuals (the optical systems) in the population;
- the maximum number of generations is the maximum allowed number of generations which will be executed before the genetic algorithm unconditionally finishes the search of the optimization space;
- the maximum radius of the curvature coefficient is the coefficient that multiplies all variable radii of the initial optical system in order to obtain the maximum values for the radius of curvature. The variable radii of curvature can be randomly selected for all optical systems in the population in the range from zero to the maximum radius of curvature;

- the maximum separation coefficient is the coefficient that multiplies all variable separations between the optical surfaces of initial optical system in order to obtain the maximum values for the separation between the optical surfaces. The variable separations between the optical surfaces can be randomly selected for all optical systems in the population in the range from zero to the maximum separation between the optical surfaces;
- the maximum aspherical coefficient is the coefficient that multiplies all variable aspherical coefficients of the initial optical system in order to obtain maximum values for the aspherical coefficients. The variable aspherical coefficients can be randomly selected for all optical system in the population in the range from zero to the maximum aspherical coefficient;
- the linear normalization parameters are as follows:
    - the maximum value for the linearly normalized merit function;
    - the minimum value for the linearly normalized merit function;
    - the linearly normalized merit function step value by which the merit function values are reduced from the maximum value to the minimum value;
- the genetic operator parameters are as follows:
    - the uniform crossover probability for selection as a genetic operator;
    - the average crossover probability for selection as a genetic operator;
    - the floating point mutation probability for selection as a genetic operator;
    - the big mutation creep probability for selection as a genetic operator;
    - the small mutation creep probability for selection as a genetic operator;
    - the floating point mutation maximum coefficient;
    - the big mutation creep maximum coefficient;
    - the small mutation creep maximum coefficient;
    - the floating point mutation probability of execution as a genetic operator;
    - the big mutation creep probability of execution as a genetic operator;
    - the small mutation creep probability of execution as a genetic operator;
- the genetic operator adaptation parameters are as follows:
    - the number of genetic operators per generation;
    - the number of executed genetic operators before the performance test;
    - the uniform crossover minimum parameter;
    - the average crossover minimum parameter;
    - the floating point mutation minimum parameter;
    - the big mutation creep minimum parameter;
    - the small mutation creep minimum parameter.

Chapter 10

Two membered evolution strategy ES EVOL implementation

10.1 General description of the two membered evolution strategy ES EVOL

It is well known that the genetic algorithms are designed as a general search method which can be, also used in the optimization of various technical systems. There is a large number of different genetic algorithms but for the optical system optimization only the adaptive steady-state genetic algorithm (ASSGA) is used because it is the only genetic algorithm which is well adapted to the optimization of complex technical systems. Contrary to the genetic algorithms, the evolution strategies are designed to be the optimization method for complex technical systems. All evolution strategies are very well suited for the optical system optimization. The following evolution strategies are implemented in the optical system optimization:

– the two membered evolution strategy ES EVOL which represents the most simple model for the evolution simulation with only two members: a parent and an offspring and with mutation as only one genetic operator;

– the multimembered evolution strategies ES GRUP, ES REKO and ES KORR which are further developments of evolution simulation from the first evolution strategy ES EVOL. The ES GRUP is developed from the ES EVOL and it introduces the population that is larger than two members but keeps mutation as an only genetic operator. The ES REKO is developed from the ES GRUP and it introduces the recombination as an genetic operator. This means that the ES REKO has two genetic operators: the mutation and the recombination. The ES KORR is a final and the most developed simulation of evolution, which has several genetic operators.

A detailed mathematical theory for both two membered and multimembered evolution strategies is given in Chapter 4. In this and next two chapters the accent will be put on the specific facts in the optical system optimization by the various evolution strategies. The basic flow chart diagram for the two membered evolution strategy ES EVOL is shown in Fig. 10.1.

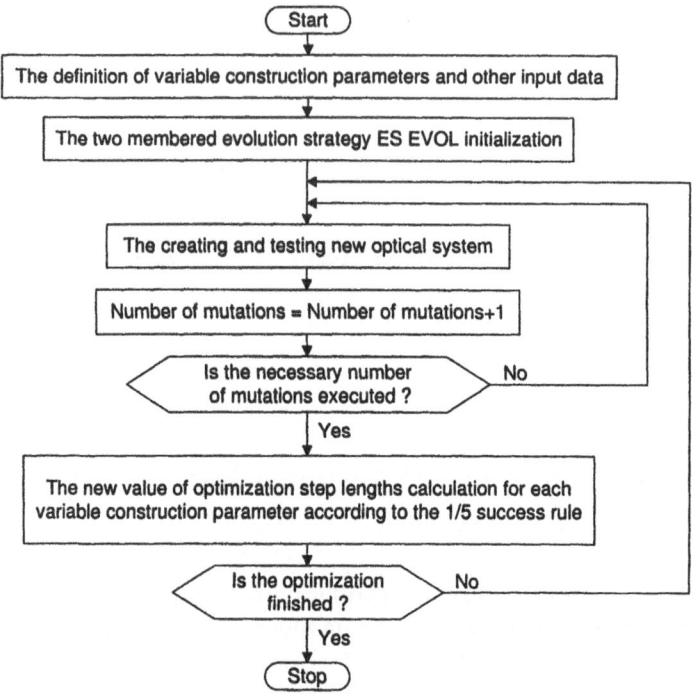

Figure 10.1. The basic flow chart diagram for the ES EVOL optimization

Fig. 10.1 shows that the optimization is started with the definition of all necessary data for the ES EVOL. First it is necessary to define all variable construction parameters like variable radii of curvature, variable separations between optical surfaces, variable glasses of which optical elements (lenses, prisms, mirrors) are made and variable aspherical coefficients if the optical system has the aspherical surfaces that can be changed in the optimization. The optical system designer must define all necessary data because only he knows the purpose and the design requirements for the optimized optical system. The definition of the variable construction parameters is actually done before the starting optimization itself. The next step is the ES EVOL initialization which consists, among other things, of detailed testing of the initial optical system. If the initial optical system fulfills all conditions, then optimization may precede and eventually find the optimum optical system with the minimum aberrations. If the initial optical system does not fulfill all the conditions, then the optimization is not possible and the search for the optical system that fulfills all the conditions is initialized. This is a quite complex part of the algorithm and it is described in detail is Section 10.2. The next step after the ES EVOL initialization is creating and testing a new optical system. The new optical system is created from the initial optical system in the first generation and from the parent optical system in all other generations by applying the mutation to the optical system. The new optical system must be tested to see if it is valid. This part of the optimization algorithm is described in detail in Section 9.3.

After creating the new optical system the number of mutations is increased and tested to see if the required number of mutations is executed for the calculation of the optimization step lengths (the standard deviations) a new values. The number of mutations that is executed before a new optimization step lengths determination is directly proportional to the number of variable construction parameters. When the required number of mutations is executed which means when the number of a new created optical systems is equal to the number of the variable construction parameters then a new values for the optimization step length can be determined according to the 1/5 rule. The mathematical theory of the 1/5 rule is given in Chapter 4.

If the number of successful mutations, i.e. the optical system created with the smaller aberrations and the merit function, is greater than 1/5 of total number of executed mutations then the optimization step length is reduced. Reducing the optimization step length is obtained by the multiplication with the control optimization variable which has the default value of 0.85 but it can be any number smaller than one. Of course the optical system designer can modify before or after optimization the default value for the control optimization variable. If the number of successful mutations is smaller than the 1/5 of the total number of executed mutations then the optimization step length is enlarged by the division with the control optimization variable. The control optimization variable can be reduced to the predefined small value which is the parameter of the optimization algorithm and must be specified before the optimization or accept the default value. After reducing the optimization step length to the predefined small value, the optimization step length becomes constant and equal to the predefined small value.

When the optimization step lengths are calculated the conditions for the ending optimization are checked. The convergence criterion for ending the optimization is the following: the optimization is finished if and only if the optical system merit function value is improved for less than sufficiently small value expressed in relative ε_{rel} or absolute ε_{abs} form in the last $10 \cdot N \cdot LR \cdot LS$ mutations:

$$\psi - \psi_{min} < \varepsilon_{abs}$$
$$\frac{|\psi - \psi_{min}|}{|\psi|} < \varepsilon_{rel}$$

$$(10.1)$$

where:

ψ is the optical system merit function value in the current iteration;

ψ_{min} is the minimum value of the optical system merit function;

ε_{obs} is the minimum value from which the optical system merit function absolute difference must be greater in the last $10 \cdot N \cdot LR \cdot LS$ mutations;

ε_{rel} is the minimum value from which the optical system merit function relative difference must be greater in the last $10 \cdot N \cdot LR \cdot LS$ mutations;

N is the number of variable construction parameters;

LR is the variable that is used in the optimization step length control. The value for the variable *LR* can be determined so that on average one success (improvement in the optical system merit function) is obtained in $5 \cdot LR$ mutations. This needs to be computed in last $10 \cdot N \cdot LR$ mutations.

LS is the variable that is used in testing the convergence criterion. Its minimum value can be 2.

10.2   ES EVOL initialization

The first step in each optical system optimization by the evolution strategies in general and the ES EVOL in particular is the initialization process which consists of:

— the important parameters definition;

— the initial optical system detailed testing;

— the decision whether the optimization of the initial optical system is possible or the optical system that fulfills all conditions must be found first and then the found optical system optimized.

The detailed flow chart diagram for the ES EVOL initialization is shown in Fig. 10.2.

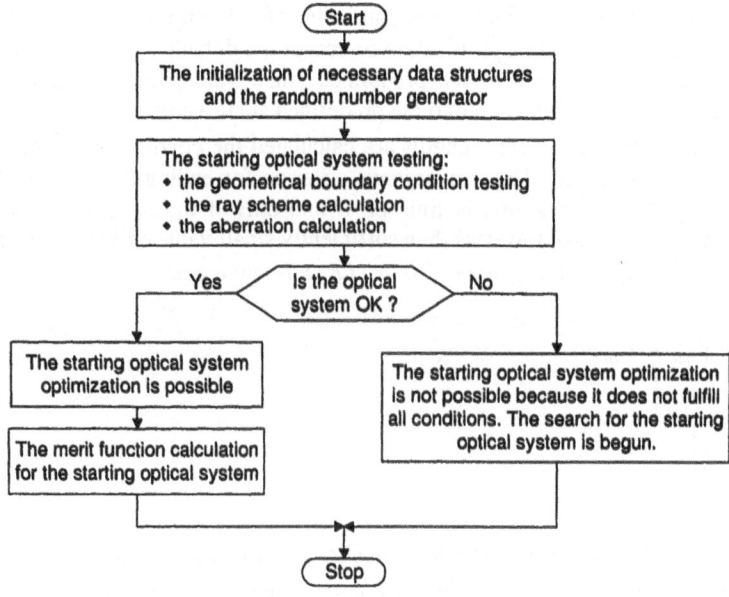

Figure 10.2. The flow chart diagram for the ES EVOL initialization

The first thing that must be done before the starting optimization algorithm is the definition of the following input parameters:

— the minimum value for the optimization step length (the standard deviation) expressed in the absolute value and relative to the values of variables. During the optimization the optimization step length can be reduced only to the optimization step length minimum value and after that the optimization step length can be constant or increase its value. The optimization step length minimum value is also the function of the optical system complexity. It is difficult to define the default value for the optimization step length minimum expressed in the absolute value and relative to the values of variables. The optical system designer can specify the optimization step length minimum value according to his experience, intuition and complexity of the optimized optical system. The usual values are between 0.001 and 0.000001.

- the minimum value for the optical system merit function difference in a generation expressed in the absolute and the relative value (ε_{aps} and ε_{rel}). There is no default value for the optical system merit function difference minimum. The optical system designer can specify it according to his experience, intuition and complexity of the optimized optical system. The usual values are between 0.001 and 0.000001. The smaller values are used for more complex optical systems with a greater number of variable construction parameters.

- the control variable in the convergence testing (LS). The optimization is terminated if the optical system merit function value has improved by less than the minimum value for the optical system merit function difference in a generation expressed in the absolute or the relative value in the last $10 \cdot N \cdot LR \cdot LS$ mutations. The default value for the control variable in the convergence testing is 2 and the optical system designer can chose any value between 2 and 5.

- the control variable in the optimization step length management (LR). The value for this control variable is determined so that on average one success (improvement in the optical system merit function) is obtained in $5 \cdot LR$ mutations. This needs to be computed in last $10 \cdot N \cdot LR$ mutations. The default value for the control variable in the optimization step length management is 1 and the optical system designer can chose any value between 1 and 5.

- the control variable in the optimization step length adjustment. With this variable the optimization step length is multiplied or divided in the process of adjusting the optimization step length during the optimization. For the default value the author accepted, Schwefel's recommendation in [21] for the control variable to be 0.85. The optical system designer can specify any value from 0.5 to 1.0 for the control variable in the optimization step length adjustment.

- the initial value for the optimization step length. For the default value the author here also accepted Schwefel's recommendation in [21] for the initial value for the optimization step length to be 1.0. The optical system designer can specify any value from 0.5 to 2.0 for the initial value for the optimization step length.

- the maximum number of allowed generations before the optimization is unconditionally finished. The maximum number of generations is the function of the optical system complexity meaning that more complex optical systems require a larger number of generations.

The most important thing in the ES EVOL initialization is a detailed testing of the initial optical system. For each optical system optimization the initial optical system must be known i.e. selected by the optical system designer. According to his experience and intuition the optical system designer selects the optical system that can represent a better (the optical system closer to the optimum) or a worse (the optical system further from the optimum) starting point. The optical system designer also selects the variable construction parameters in the optical system. By changing this variable construction parameters the optimization algorithm can find a better optical system with smaller aberrations and the merit function. The designer of optical systems can usually make the following construction parameters variable:

− the radii of curvature;

− the separations between the optical surfaces;

− the glasses of which the optical elements (lenses, prisms and mirrors) are made;

− the aspherical surface coefficients if the optical system contains aspherical surfaces.

It is very important to notice that the optical system designer can make variable only some construction parameters and never all parameters. The reason is that in the optimization, a type of optical system is not allowed to be changed. This means that the arrangement of the optical components must be constant i.e. if the doublet is optimized it will be also the doublet at the end of optimization but it will have some of the radii of curvature changed, as well as the separations between the optical surfaces and the glasses which will allow it to have smaller aberrations and the merit function. It is important to notice that the doublet will never transform to, e.g., the triplet during the optimization nor the converging and diverging lenses will change their places.

Each optical system is defined with the record that contains the following basic construction data:

− the radii of the curvature for all optical surfaces;

− the distances between the optical surfaces;

− the glasses of which the optical elements (lenses, prisms) are made. The glasses are represented by the glass name from the appropriate glass manufacturer, the index of the refraction for three spectral lines and the dispersion. The index of refraction and the dispersion are used for the optical system calculation.

− the clear radii for all optical surfaces;

− the aspherical surface coefficients for all aspherical optical surfaces if there are any aspherical surfaces in the optical system;

In order to define the optical system completely it is necessary to, define a record with the ray data besides the basic record with the construction data, i.e. the coordinates of the selected set of rays on the entrance reference plane. This set of rays simulates the passage of the light through the optical system and it is used for the calculation of all optical system properties like aberrations, the optical path difference, the spot diagram, the modulation transfer function and the merit function.

The initial optical system must be tested to see whether all conditions for the optimization are fulfilled. The process of the optical system testing consists of the following:

- the geometrical boundary condition testing which consists of the following:
  - the optical system must have all radii and separations physically realizable;
  - in the optical system all axial and peripheral lens thicknesses must be greater than the predefined standard values;
  - in the optical system all clear apertures must be smaller than the predefined values;
  - the predefined ray scheme must be possible to be calculated for the initial optical system;

  It is important to notice that the optical system designer must select or accept default values for all these predefined quantities;
- the paraxial values calculation;
- the aberrations and the merit function calculation.

If any of these conditions is not fulfilled then the initial optical system cannot become the parent for the first generation and the optimization cannot start. The optimization algorithm then starts the search for the optical system that fulfills all conditions. In this search the same optimization algorithm, the two membered evolution strategy ES EVOL, is used. Instead of the classical merit function that is based on aberrations, the auxiliary merit function that represents the deviation from the boundary conditions is formed. The auxiliary merit function is minimized in order to fulfill all necessary conditions. When the optical system that fulfills all conditions is found then the real optimization with the aberration minimization can be started.

If all these conditions are fulfilled then the initial optical system can become the parent for the first generation and the optimization can start. In the optimization each optical system is characterized by the merit function, which is necessary to be calculated. The merit function is inherited from the classical damped least squares optimization and it is described in Chapter 8.

10.3 Creating and testing a new optical system in the ES EVOL optimization

The main part of the optical system optimization by the evolution strategies in general and the ES EVOL in particular is creating and testing the optical system. By creating a new optical systems the optimization algorithm can search the optimization space and find good optical systems that can represent a possible solution to the problem. The detailed flow chart diagram for optical system creating and testing in the two membered evolution strategy ES EVOL optimization is shown in Fig. 10.3.

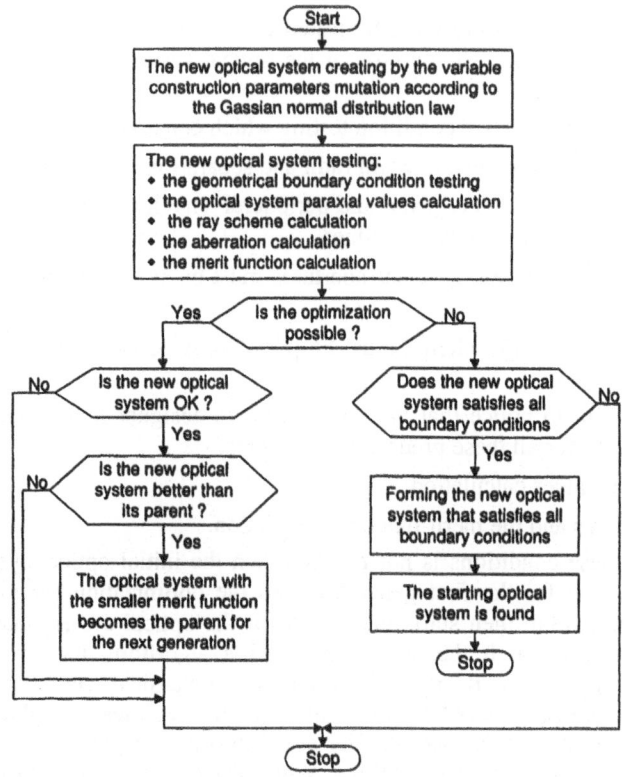

Figure 10.3. The flow chart diagram for the creating and testing optical system in the ES EVOL optimization

From the flow chart diagram shown in Fig. 10.3 it can be seen that the creation of a new optical system is done by the variable construction parameter mutation. The mutation is done according to the Gaussian normal distribution law. The algorithm for the mutation as follows:

A random number is generated according to the Gaussian normal distribution law. The transformation of Box and Muller described in Chapter 4 is used to obtain the normally distributed random numbers starting from the uniformly distributed random numbers. This random number is multiplied with the optimization step length (the standard deviation) for the corresponding variable construction parameter. The obtained number from the random number and the optimization step length multiplication are added to the variable construction parameter and a new value for the parameter is formed. Most values for the variable construction parameter will be close to the starting values of the variable construction parameters because the normal distribution with the mean value zero and the standard deviation equal to the optimization step length is used. Some new values for the variable construction parameter will be equally distant from the positive and the negative side of the variable construction parameter starting value. This is especially true in the beginning of the optimization when the standard deviation value is quite large, which enables the search of the large optimization space.

The basic supposition in the optical system optimization is to try to improve the selected optical system and not to form a new optical system type. This means that if a classical cemented doublet is optimized then after the optimization it will be also a cemented doublet not, for example, triplet, but it will have the radii of curvature changed as well as the separations between the optical surfaces and/or glasses in order to improve the aberrations. It is well known that the classical cemented doublet consists of only two lenses: the first is converging and the second is diverging. During the optimization it is never possible for converging and diverging lenses to change their places. Any mutation of the variable construction parameter that changes the optical system type is not allowed and this mutation is rejected and a new mutation of the variable construction parameter is created. The mutation of the variable construction parameter is created and tested in the loop until the mutation fulfills all the necessary conditions. The mutations of the variable construction parameters must also fulfill certain geometrical conditions such as:

– the absolute value of the radii must be greater than the clear aperture;
– the lens thicknesses must be physically realizable i.e. must be greater than the minimum axial lens thickness.

All these geometrical conditions must be checked as soon as a new variable construction parameter is created by the mutation. If any of the geometrical conditions is not fulfilled then this a new variable construction parameter is rejected and a new one is generated. A new variable construction parameter creation by the mutation and testing is done in the loop until all conditions are fulfilled.

After all a new values for the variable construction parameters are generated by the mutation and tested, the new optical system is formed. The new optical system must be tested with the following tests:

– the paraxial value calculation;
– the ray scheme calculation;
– the calculation of aberrations and merit function.

The next step in the optimization algorithm is the decision whether the optical system optimization is possible or not. If the optimization is possible then the search for the optical system with the smaller aberrations and the merit function is continued and if the optimization is not possible then the search for the optical system fulfilling all the conditions is continued.

If the optimization is possible then a new optical system is tested and if it fulfills all conditions the parent for a new generation is selected. As opposed to the genetic algorithms where the parents were selected randomly by the roulette wheel simulation in evolution strategies, the parents are selected deterministically. The optical system merit functions of two individuals in the population (the parent and the offspring) are compared and the better individual with the smaller merit function is selected to be the parent for the next generation.

If the optimization is not possible then the search for the initial optical system that fulfills all the conditions is conducted. The first thing that must be done is to test whether a new optical system fulfills all the conditions. In order to find the initial optical system fulfilling all the conditions, the auxiliary merit function, containing the sum of the boundary condition deviations, is formed. The same optimization algorithm, the two membered evolution strategy ES EVOL, is trying to minimize the auxiliary merit function i.e. the optimization algorithm is trying to make the auxiliary merit function equal to the zero which means that there are no boundary condition violations. If the optical system fulfilling all the conditions is found then the search for the initial optical system is finished and this newfound optical system becomes the initial optical system for the real optimization where the aberrations are minimized. If the new formed optical system does not fulfill all the conditions then the auxiliary merit function is calculated for this new optical system. The auxiliary merit function of two optical systems in the population (the parent and the offspring) is compared and the optical system with the smaller auxiliary merit function is selected for the parent in the next generation.

Chapter 11

Multimembered evolution strategies
ES GRUP and ES REKO implementation

11.1   General description of the multimembered evolution strategies ES GRUP and ES REKO

The two membered evolution strategy represents the simplest model of the evolution simulation with only two members a parent and an offspring and only one genetic operator – the mutation. The multimembered evolution strategies represent a more complex model of the evolution simulation. In this chapter the implementation of two variants of the multimembered evolution strategies is described:

– the ES GRUP which is the multimembered evolution strategy with the mutation according to the Gaussian normal distribution law as the only genetic operator;

– the ES REKO which is the multimembered evolution strategy with two genetic operators: the variable construction parameter mutation according to the Gaussian normal distribution law and the optimization step length recombination which is in fact the calculation of the arithmetical mean value of optimization step lengths.

A detailed mathematical theory for both two membered and multimembered evolution strategies is given in Chapter 4. In this chapter the stress will be put on specific facts in the optical system optimization by the multimembered evolution strategies ES GRUP and ES REKO.

It is important to notice that the multimembered evolution strategies ES GRUP and ES REKO are developed from the two membered evolution strategy ES EVOL and some parts of the optimization algorithm for the ES EVOL, ES GRUP and ES REKO are identical. The basic flow chart diagram for the ES GRUP and ES REKO is shown in Fig. 11.1.

From the Fig. 11.1 can be seen that the optimization is started with the definition of all the necessary data for the multimembered evolution strategies ES GRUP or ES REKO. First it is necessary to define all variable construction parameters like variable radii of curvature, variable separations between the optical surfaces, variable glasses that optical elements (lenses, prisms, mirrors) are made of and variable aspherical coefficients if the optical system has the aspherical surfaces that can be changed in the optimization. The optical system designer must define all necessary data because only he knows the purpose and the design requirements for the optimized optical system. The definition of the variable construction parameters is actually done before starting the optimization itself. The next step is the ES GRUP or the ES REKO initialization consisting among other things of detailed testing of the initial optical system. If the initial optical system fulfills all conditions then the optimization may precede and eventually find the optimum optical system with the minimum aberrations. If the initial optical system does not fulfill all the conditions then the optimization is not possible and the search for the optical system that fulfills all the conditions is initialized. This is quite a complex part of the algorithm and it is described in detail in Section 11.2.

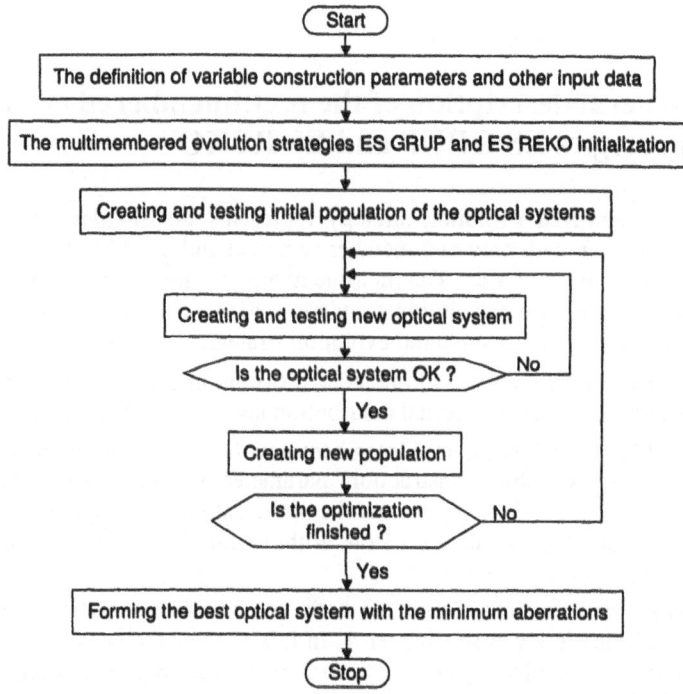

Figure 11.1. The basic flow chart diagram for the ES GRUP and the ES REKO optimization

The next step after the multimembered evolution strategies initialization is creating and testing the initial population. The initial population is formed of the starting optical system by changing the variable construction parameters with the mutation according to the Gaussian normal distribution law. The first individual in the initial population is always the starting optical system. Each new created optical system is tested and only if it fulfills all conditions a new optical system is accepted in the population. The initial population creating and testing is a rather complex process and it is described in detail in Section 11.3. After the successful generation of the initial population the optimization algorithm can be started. The search of the optimization space is performed by creating a new optical systems, which is done differently in the ES GRUP and the ES REKO evolution strategy. In the evolution strategy ES GRUP the new optical system creating is done only by the mutation of the variable construction parameters and in the evolution strategy ES REKO the new optical system creating is done by the mutation of the variable construction parameters and the recombination of the optimization step lengths (standard deviations).

Each new optical system must be tested and if it fulfills all conditions then it can be accepted to the population, otherwise it is rejected. By creating a new optical systems a new population is created and a new parents for the next generation can be selected. The optimization algorithm search the optimization space in order to find the minimum i.e. the optical system with the minimum aberrations and the merit function by creating a new offspring and selecting parents for the next generation.

When the whole new population is formed and the best optical system with the minimum merit function is selected then the conditions for ending the optimization are checked. The convergence criterion for ending the optimization is the same as for the two membered evolution strategy ES EVOL and it is described in Chapter 10.

11.2   Evolution strategies initialization

The first step in every optical system optimization by the evolution strategies in general and the multimembered evolution strategies ES GRUP or ES REKO in particular is the initialization process which consists of important parameters definition and the detailed testing of the initial optical system and the decision whether the optimization of the initial optical system is possible or the optical system that fulfills all conditions must be found in the first place and then the found optical system optimized. The detailed flow chart diagram for the multimembered evolution strategies ES GRUP and ES REKO initialization is shown in Fig. 11.2.

The first thing that must be done before starting the optimization algorithm is the definition of the input parameters. Only the input parameters different from the two membered evolution strategy ES EVOL will be described here. The detailed description of all other input parameters is given in Chapter 10. The input paramters specific for the multimembered evolution strategies ES GRUP and ES REKO are:

– the number of parents (μ) in the population. The number of parents must be greater than 1. The default value for the number of parents is 10 but the optical system designer is free to change this value.

– the number of offspring (λ) which are created during the generation and which must be greater than or equal $\lambda \geq 6 \cdot \mu$. The default value for the number of offspring is 100 but the optical system designer is free to change this value.

For each optical system optimization the initial optical system must be known i.e. selected by the optical system designer. The whole description of the initial optical system selection and the definition of the variable construction parameters is given in Chapter 10. The initial optical system must be tested to be seen whether the all conditions for the optimization are fulfilled. The process of the optical system testing is the same for all evolution strategies and it is described in Chapter 10.

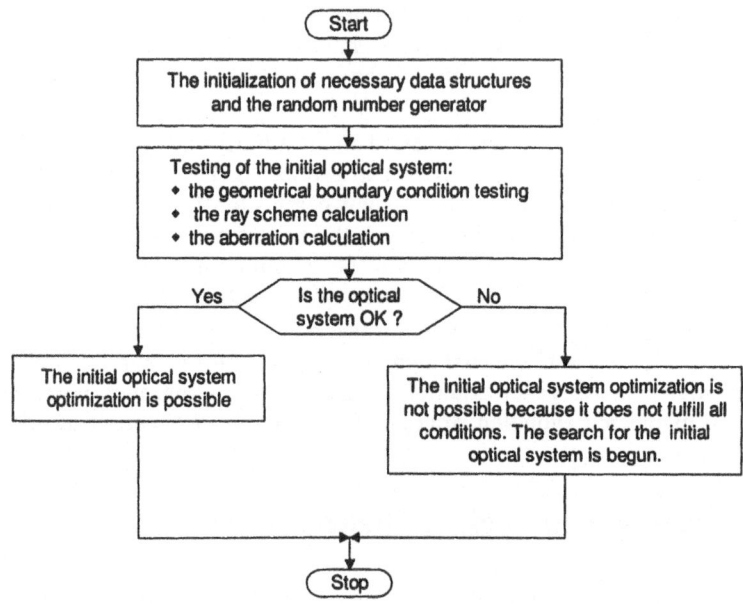

Figure 11.2. The flow chart diagram for the ES GRUP and the ES REKO initialization

If any of conditions in the initial optical system testing are not fulfilled then the initial optical system cannot become the parent for the first generation and the optimization cannot start. The optimization algorithm then starts the search for the optical system that fulfills all conditions. For this search the same optimization algorithm the multimembered evolution strategies ES GRUP or ES REKO is used. Instead of the classical merit function that is based on aberrations, the auxiliary merit function representing the deviation from the boundary conditions is formed. The auxiliary merit function is minimized in order to fulfill all necessary conditions. When the optical system that fulfills all conditions is found then the real optimization with the aberration minimization can be started.

11.3 Creating and testing initial population in the multimembered evolution strategies ES GRUP and ES REKO

When the initial optical system is tested then the initial population can be generated from the initial optical system by using only the mutation of the variable construction parameters in the ES GRUP optimization or the mutation of variable construction parameters and the recombination of the optimization step lengths (the standard deviations) in the ES REKO optimization. The detailed flow chart diagram for the creating and testing initial population in the multimembered evolution strategies ES GRUP and ES REKO is shown in Fig. 11.3.

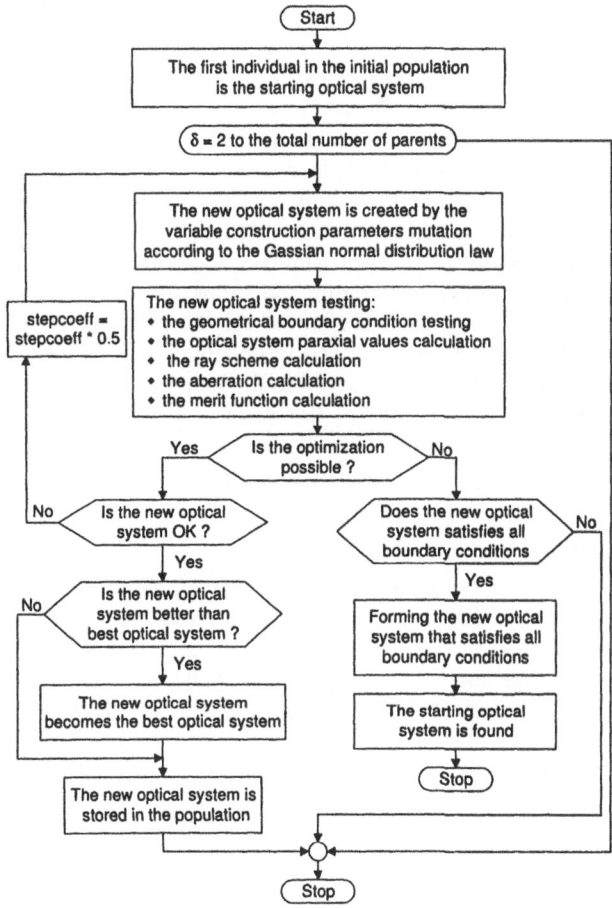

Figure 11.3. The flow chart diagram for the creating and testing initial population in the multimembered evolution strategies ES GRUP and the ES REKO

The first thing that is necessary to do in the creating and testing of the initial population is to copy the starting optical system into the initial population. The rest of the initial population is formed with the optical systems having slight variations of the variable construction parameters according to the Gaussian normal distribution law. Each new created optical system for the initial population must fulfill some conditions.

The basic supposition in the optical system optimization is to try to improve the selected optical system and not to form a new optical system type. This means that if the classical cemented doublet is optimized then after the optimization it will be also the cemented doublet not, for example, the triplet, but it will have changed radii of curvature and/or separations between the optical surfaces and/or glasses in order to improve the aberrations. It is well known that the classical cemented doublet consists of only two lenses the first is converging and the second is diverging. During the optimization it is never possible that converging and diverging lenses change their places.

Any mutation of the variable construction parameter that changes the optical system type is not allowed and this mutation is rejected and a new mutation of the variable construction parameter is created. Any rejected mutation of the variable construction parameter means that the mutation is too large and that it must be reduced in order to be successful. This reduction of the variable construction parameter mutation is done by the correction coefficient which has the value 0.5 and multiplies the product of the optimization step length for this variable construction parameter and the generated random number according the Gaussian normal distribution law. The process of creating mutations of the variable construction parameters is described in the next section where the process of the creating and testing the new optical system is described in details. The mutation of the variable construction parameter is created and tested in the loop until the mutation fulfills all necessary conditions. The mutations of the variable construction parameters also must fulfill certain geometrical conditions such as:

– the absolute value of the radii must be greater than the clear aperture;

the lens thickness must be physically realizable i.e. must be greater than the minimum axial lens thickness.

All these geometrical conditions must be checked as soon as a new variable construction parameter is created by the mutation. If any of the geometrical conditions is not fulfilled then this new variable construction parameter is rejected and a new one is generated. Each rejected mutation must be reduced by multiplication with the correction coefficient as described above. A new variable construction parameter creating and testing is done in the loop until all the conditions are fulfilled.

After all new values for the variable construction parameters are generated by the mutation and tested a new optical system is formed. A new optical system must be tested by the following tests:

– the paraxial value calculation;

– the ray scheme calculation;

– the calculation of aberrations and merit function.

If a new optical system does not pass all the tests, it is rejected and a new one is created in the same way as the rejected optical system but the mutations that create the new optical system are reduced by the multiplication with the correction coefficient as described above.

If a new optical system does pass all the tests it is accepted to the population and its merit function is calculated. The merit function is inherited from the classical damped least squares optimization and it is described in Chapter 8.

In optimization it is always very important to know which optical system is the best, i.e. which optical system has the minimum aberrations. To be able to know which optical system is the best each new optical system must be compared with the best optical system. The comparison of the optical systems is in fact the comparison of its merit functions. If a new optical system has the merit function less than the best optical system then it becomes the best optical system.

If the optimization is not possible i.e. the initial optical system does not fulfill all conditions, then a new optical system fulfilling all conditions, must be found. In order to find the optical system that fulfills all conditions, each new created optical system is tested to be seen if it fulfills all conditions. If a new optical system does so, then the search for the optical system fulfilling all conditions is finished and the optical system becomes the initial optical system for the real optimization in which the aberrations are minimized.

11.4 Creating and testing a new optical system in the multimembered evolution strategies ES GRUP and ES REKO

The main part of the optical system optimization by the evolution strategies in general and the multimembered evolution strategies ES GRUP and ES REKO in particular is creating and testing the optical system. By creating a new optical systems the optimization algorithm can search the optimization space and find good optical systems that can represent solution to the problem. The detailed flow chart diagram for creating and testing the optical system in the multimembered evolution strategies ES GRUP and ES REKO is shown in Fig. 11.4.

The first thing in creating and testing the optical system is a decision whether the optimization is done with the ES GRUP algorithm or the ES REKO algorithm. This is done because the ES GRUP and ES REKO have different algorithms for creating a new optical systems.

When the ES GRUP algorithm is used a new optical system is created by using the mutations of the variable construction parameter, which is very similar to the mutation in the two membered evolution strategy ES EVOL and will not be described here because the whole description is given in Chapter 10.

When the ES REKO algorithm is used, a new optical system is created by using two genetic operators: the mutations of the variable construction parameters and the recombination of the optimization step lengths (the standard deviations) for each variable construction parameter. The process of creating a new variable construction parameter is the following one:

The first step is the recombination of the optimization step lengths of two randomly chosen parents from the population. The recombination is the intermediate recombination, meaning that a new value for the optimization step length of the variable construction parameter is the arithmetical mean value of the selected optimization step lengths. When a new optimization step length is created by recombination, the optimization algorithm can proceed with creating a new variable construction parameter by the mutation. The process of creating a new variable construction parameter is the same as in the ES GRUP optimization and the ES EVOL optimization. The description is given in Chapter 10.

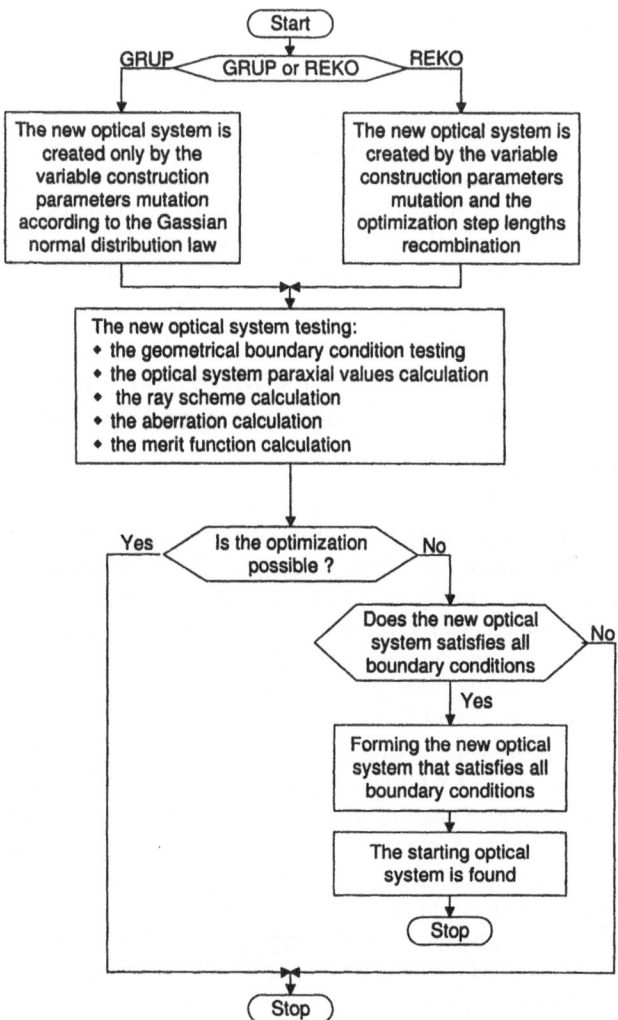

Figure 11.4. The flow chart diagram for creating and testing the new optical system in the multimembered evolution strategies ES GRUP and ES REKO

It is important to notice that in creating the new optical system by the ES GRUP algorithm only one parent is randomly selected from the population and it is used for creating the offspring by the mutation of the variable construction parameters. When a new optical system is created by the ES REKO algorithm three parents must be randomly selected from the population. Two parents are used for the recombination of the optimization step lengths for each variable construction parameter and one parent is used for creating the offspring by the mutation of the variable construction parameters.

When a new optical system is created it must be tested to be seen whether all conditions are fulfilled. The detailed description of the optical system testing is given in Chapter 10 and it will not be described here in details.

At the end after the new optical system creation and testing the optimization algorithm must decide whether the optimization is possible or not. If the optimization is possible then a new optical system is used in searching the optimization space and finding the optical system with the minimum aberrations. If the optimization is not possible then a new optical system is used in the search for the optical system that fulfills all conditions. This is done by testing each new created optical system to see if it fulfills all conditions. If the new optical system fulfills all conditions then the search for the optical system that fulfills all conditions is finished and that optical system becomes the initial optical system for the real optimization in which the aberrations are minimized.

11.5 Creating a new population in the multimembered evolution strategies ES GRUP and ES REKO

After creating the new optical system that satisfies all conditions it is placed in a new population. The process of creating the new population is rather complex because the number of a new created offspring is larger than the number of parents. If each new created offspring is stored in the population then it will be the uneconomical use of computer memory. Thus only the best offspring i.e. the optical systems with the smallest aberrations and the merit function are stored in the population. The number of stored optical systems is equal to the number of parents for the next generation. The detailed flow chart diagram for creating the new population is shown in Fig. 11.5.

The algorithm for storing the best offspring in the population is follows:

When a new optical system that fulfills all conditions is created, the number of correct a new offspring is increased. If the number of correct a new offspring is smaller than the number of parents in the next generation the offspring is stored without testing. If the number of correct a new offspring is greater than the number of parents for the next generation then the offspring is not stored directly because all storage places are already occupied.

To be stored in the population a new optical system must have the smaller merit function than the worst optical system in the population with the biggest merit function. After the new optical system with smaller merit function is stored the whole population must be sorted and the worst optical system with the biggest merit function must be found. When all new created offspring are tested and stored in the population then it is possible to find the best optical system with the smallest merit function.

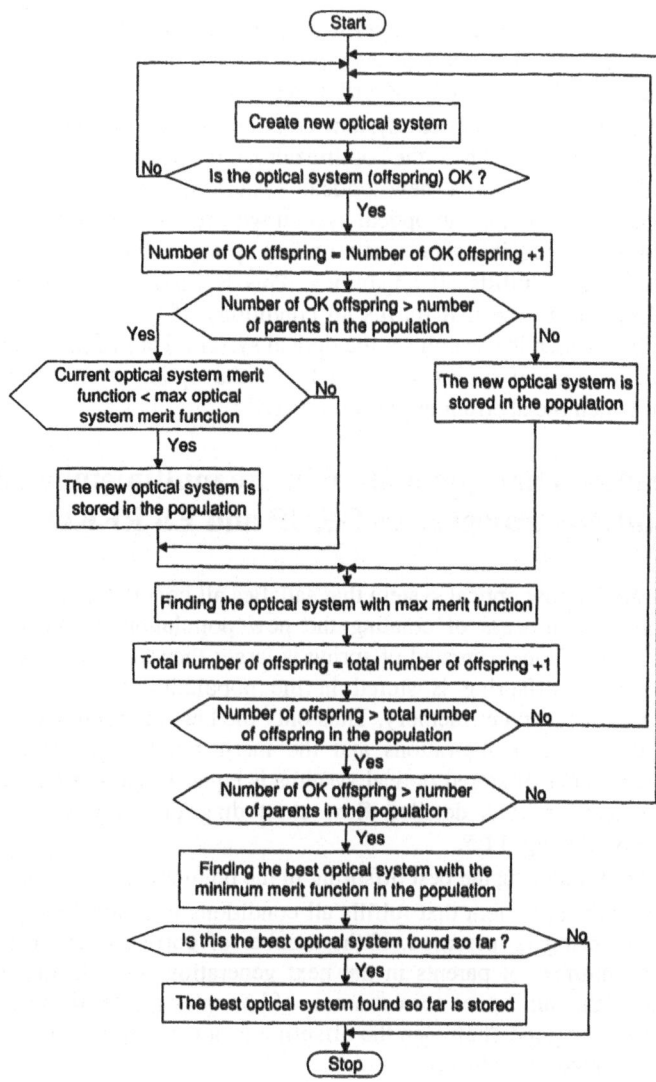

Figure 11.5. The flow chart for creating the new population in the multimembered evolution strategies ES GRUP and ES REKO

Chapter 12

Multimembered evolution strategy ES KORR implementation

12.1 General description of the multimembered evolution strategy ES KORR

The optimization method called the multimembered evolution strategy ES KORR represents a synthesis of all developed evolution strategies and their further development. The multimembered evolution strategy ES KORR is the result of the research Schwefel conducted in order to improve existing multimembered evolution strategies ES GRUP and ES REKO. Detailed mathematical theory of the general evolution strategies (two membered and multimembered) is given in Chapter 4 and here will be described only unique improvements of the ES KORR strategy compared to the rest of multimembered evolution strategies and specific data for the optical system optimization.

The multimembered evolution strategy ES KORR contains both the plus and the coma strategy as a strategy for selection. The ES KORR optimization has five different genetic operators for the recombination and linearly correlated mutations. The important improvement which is introduced by the ES KORR optimization is the self-adaptation of the optimization step lengths (the standard deviations), which means that the optimization step lengths for each variable construction parameter is separately adjusted. As a result of this self adjustment, the automatic adjustment of the variable construction parameters is obtained which in some cases contributes to the substantial improvement in the speed of convergence to the global minimum. The optimization step lengths at the two membered evolution strategy ES EVOL can be represented as concentric circles with equal probability. The optimization step lengths at the multimembered evolution strategy ES GRUP and REKO can be represented as concentric ellipses, which can be extended or contracted along the principal axis. At the ES KORR optimization these concentric ellipses representing the optimization step lengths are transformed to the hyperellipsoids, which can be expanded or contracted along all directions following the n-dimensional normal distribution of the set of n random components z_i for $i = 1(1)n$.

$$\omega(z) = \frac{1}{(2\pi)^{\frac{n}{2}} \cdot \prod_{i=1}^{n} \sigma_i} \cdot \exp\left(-\frac{1}{2}\sum_{i=1}^{n}\left(\frac{z_i}{\sigma_i}\right)^2\right) \tag{12.1}$$

where σ_i is the optimization step length (the standard deviation).

In this case the hyperellipsoid can extend or contract along the coordinate axis or even to rotate in order to obtain the most convenient position in the optimization space. During the hyperellipsoid rotation the random components z_i become mutually dependent or correlated. The simplest kind of random components correlation is linear which is only case to yield hiperellipsoids as surfaces of constant optimization step length probability.

In the most general case the number of hyperellipsoid rotation angles n_p which can take all values between the $0°$ and $360°$ is:

$$n_p = \frac{n}{2} \cdot (n-1) \tag{12.2}$$

where n is the number of the variable construction parameters. The total number of variable construction parameters that can be specified and changed with the mutation and the recombination in the multimembered evolution strategy ES KORR is the sum of the rotation angles and the optimization step lengths:

$$n_{tot} = n_p + n_s = \frac{n}{2} \cdot (n+1) \tag{12.3}$$

where n_s is the total number of the optimization step lengths.

In the simplest case, which can be represented as a simple ellipse ($n = n_s = 2$, $n_p = 1$) the coordinate transformation for the rotation can be defined as [21]:

$$\begin{aligned}
\Delta x_1 &= \Delta x_1' \cdot \cos\alpha - \Delta x_2' \cdot \sin\alpha \\
\Delta x_2 &= \Delta x_1' \cdot \sin\alpha + \Delta x_2' \cdot \cos\alpha
\end{aligned} \tag{12.4}$$

For $n = n_s = 3$ three concentric rotations are needed to be made:

− in the ($\Delta x_1, \Delta x_2$) plane through an angle α_1 ;
− in the ($\Delta x_1', \Delta x_2'$) plane through an angle α_2 ;
− in the ($\Delta x_2'', \Delta x_3''$) plane through an angle α_3 .

Starting from the uncorrelated random changes $\Delta x_1''', \Delta x_2''', \Delta x_3'''$ these rotations have to be made in reverse order. Thus also in the general case with the $\dfrac{(n-1)\cdot n}{2}$ rotations each one only involves two coordinates as it is shown for the case of three rotations.

The multimembered evolution strategy ES KORR optimization method has the following possibilities:

- the variable optimization parameters can be:
    - construction parameters (the radii of curvature, the separations between the optical surfaces, the glasses and the aspherical coefficients);
    - the optimization step lengths (the standard deviations);
    - the mutation hyperellipsoid rotation angles.
- two ways of selecting parents for the next generation:
    - the plus strategy where the parents for the next generation are selected between the parents of this generation and all their offspring;
    - the coma strategy where the parents for the next generation are selected only between the offspring that are created in this generation.
- five different types of genetic operators:
    - the mutation according to the Gaussian normal distribution law which is a pure mutation of the variable construction parameters without the recombination of optimization step lengths;
    - the discrete recombination of the parent pairs which means that two parents are randomly selected from the whole population and the offspring gets the randomly selected variable optimization parameters from the first or the second parent;
    - the intermediary recombination of parent pairs which means that two parents are randomly selected from the whole population and the variable optimization parameters of offspring are the arithmetical mean value of the corresponding variable optimization parameters from the selected parents;
    - the discrete recombination of all parents in the population which means that for each variable optimization parameter of offspring two parents are randomly selected and the variable optimization parameter is randomly selected between these two parents;
    - the intermediary recombination of all parents in the population which means that for each variable optimization parameter of offspring two parents are randomly selected and the variable optimization parameter is equal to the arithmetical mean value of the corresponding variable optimization parameters from the selected parents.

The important thing is to notice that in the discrete recombination of all parents in the population and the intermediary recombination of all parents in the population for each variable optimization parameter two new parents are randomly selected from the whole population. In that way the whole population is contributing to the creation of a new offspring. It is contrary to the classical recombination (the discrete recombination of the parent pairs and the intermediary recombination of parent pairs) where only two randomly selected parents contribute to the creation of a new offspring.

The basic flow chart diagram for the multimembered evolution strategy ES KORR is shown in Fig. 12.1.

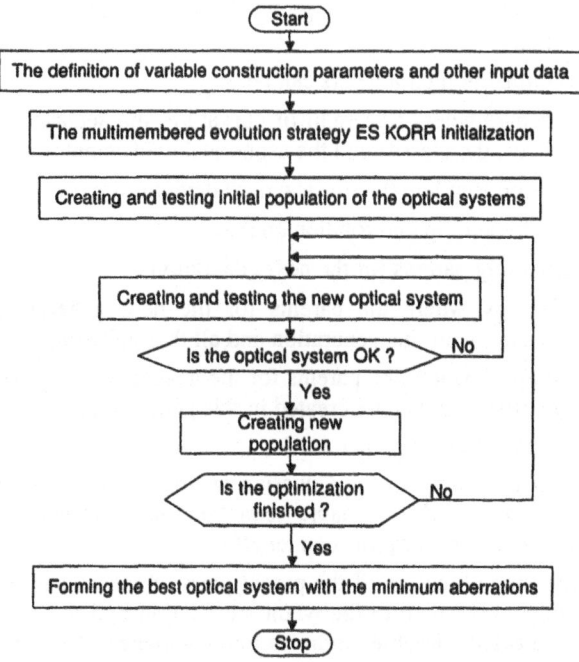

Figure 12.1. The flow chart diagram for the multimembered evolution strategy ES KORR

Fig. 12.1 shows that the optimization is started with the definition of all necessary data for the multimembered evolution strategy ES KORR. First, it is necessary to define all variable construction parameters like variable radii of curvature, variable separations between the optical surfaces, variable glasses from which optical elements (lenses, prisms, mirrors) are made and variable aspherical coefficients if the optical system has the aspherical surfaces that can be changed in the optimization. The optical system designer must define all necessary data because only he knows the purpose and the design requirements for the optimized optical system. The definition of the variable construction parameters is actually done before starting the optimization itself. The next step is the ES KORR initialization, which consists among other things, of detailed testing of the initial optical system. If the initial optical system fulfills all the conditions then the optimization may precede and eventually find the optimum optical system with the minimum aberrations. If the initial optical system does not fulfill all the conditions then the optimization is not possible and the search for the optical system that fulfills all the conditions is initialized. This is quite a complex part of the algorithm and it is described in detail in Section 12.2.

The next step after the multimembered evolution strategies initialization is creating and testing the initial population. The initial population is formed from the initial optical system by small changes in the variable construction parameters. The first individual in the initial population is always the starting optical system. Each new optical system is tested and only if it fulfills all the conditions a new optical system is accepted in the population. The initial population creating and testing is rather a complex process and it is described in detail in Section 11.3. After successful generation of the initial population the optimization algorithm can be started. The search of the optimization space is performed by creating a new optical system, which can be done by applying one of five different genetic operators to the selected parents.

Each new optical system must be tested and if it fulfills all the conditions then it can be accepted to the population otherwise it is rejected. By creating a new optical system new population is created and new parents for the next generation can be selected. The optimization algorithm search the optimization space in order to find the minimum i.e. the optical system with the minimum aberrations and the merit function by creating a new offspring and selecting parents for the next generation.

When the whole new population is formed and the best optical system with the minimum merit function is selected then the conditions for ending optimization are checked. The convergence criterion for ending optimization is the same as for all evolution strategies and it is described in Chapter 10.

12.2   ES KORR initialization

The first step in every optical system optimization by the evolution strategies in general and the multimembered evolution strategy ES KORR in particular is the initialization process which consists of important parameters definition and the detailed testing of the initial optical system and the decision whether the optimization of the initial optical system is possible or the optical system that fulfills all the conditions must be found in the first place and then the found optical system optimized. The detailed flow chart diagram for the multimembered evolution strategy ES KORR initialization is shown in Fig. 12.2.

The first thing that must be done before starting the optimization algorithm is the definition of the input parameters. Because of the multimembered evolution strategy ES KORR larger possibilities it is normal that this optimization algorithm has large number of input parameters. The input parameters are the following:

- the number of parents (μ) in the population. The number of parents must be greater than 1. The default value for the number of parents is 10 but the optical system designer is free to change this value.

- the number of offspring (λ) which are created during the generation and which must be greater than or equal to $\lambda \geq 6 \cdot \mu$. The default value for the number of offspring is 100 but the optical system designer is free to change this value.

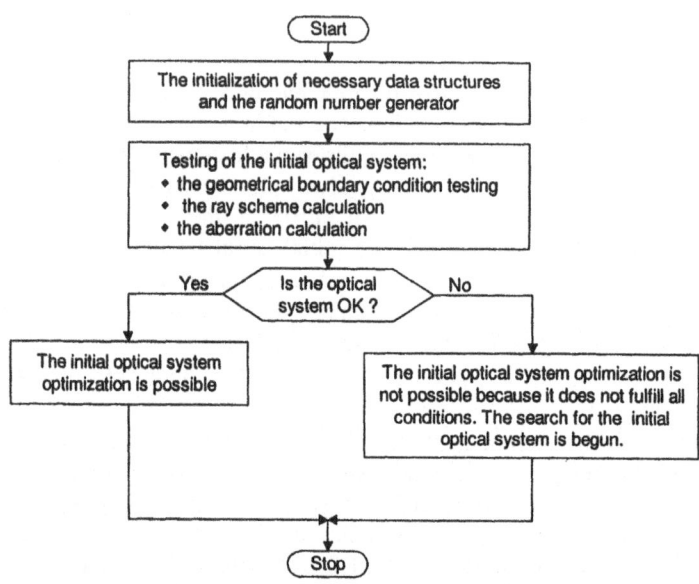

Figure 12.2. The flow chart diagram for the multimembered evolution strategy ES KORR initialization

- the selection between the coma and the plus strategy. The sefault value is the coma strategy because Schwefel in [21] shows that the coma strategy is a more successful strategy. The selection between the coma and the plus strategy is done with the logical variable.
- the parameter which defines the number of executed generations after which the checking of convergence criterion is done. The default value is 1, meaning that the convergence criterion is checked after each generation.
- the parameter defining the genetic operator for each variable optimization parameter. The variable construction parameters, the optimization step lengths and the rotation angles of the mutation hyperellipsoid can have different genetic operators. The default value is that all variable optimization parameters have the same genetic operator which is the intermediary recombination of all parents in pairs.
- the number of variable optimization step lengths. The default value is that the number of variable optimization step lengths is equal to the number of variable construction parameters. In some cases Schwelel in [21] shows that it is recommended to reduce the number of optimization step lengths.
- the maximum number of allowed generations before the optimization is unconditionally finished. The maximum number of generations is the function of the optical system complexity meaning that more complex optical systems require a larger number of generations.

- the minimum value for the optimization step length expressed in the absolute value and relative to the values of variables. This input parameter is the same for all evolution strategies and is described in Chapter 10.
- the minimum value for the optical system merit function difference in a generation expressed in the absolute and the relative value. This input parameter is the same for all evolution strategies and is described in Chapter 10.
- the initial value for the optimization step length. The author accepted, for the default value, the Schwefel's recommendation in [21] for the initial value for the optimization step length to be 1.0. According to Schwefel in [21] this is the optimum value when the number of parents is 10 and the number of offspring is 100. The optical system designer can specify any value from 0.5 to 2.0 for the initial value for the optimization step length.
- the initial value for the rotation angles in the mutation hyperellipsoid. The default value is 0. The initial value is arbitrary but because of self-adaptation they quickly obtain required values.

For each optical system optimization the initial optical system must be known i.e. selected by the optical system designer. The whole description of the initial optical system selection and the definition of the variable construction parameters is given in Chapter 10. The initial optical system must be tested to see whether all the conditions for the optimization are fulfilled. The process of the optical system testing is the same for all evolution strategies and it is described in Chapter 10.

If any of conditions in the initial optical system testing is not fulfilled then the initial optical system cannot become the parent for the first generation and the optimization cannot start. The optimization algorithm then starts the search for the optical system that fulfills all conditions. For this search the same multimembered evolution strategy ES KORR optimization algorithm is used. Instead of the classical merit function that is based on aberrations the auxiliary merit function that represents the deviation from the boundary conditions is formed. The auxiliary merit function is minimized in order to fulfill all necessary conditions. When the optical system that fulfills all conditions is found then the real optimization with the aberration minimization can be started.

12.3 Creating and testing the initial population in the multimembered evolution strategy ES KORR

The first step at each multimembered evolution strategy (ES GRUP, ES REKO and ES KORR) is the creating and testing of the initial population. The way of creating this initial population is specific for each optimization method. The multimembered evolution strategy ES KORR has five different genetic operators, which enables the great variety in the individuals of the initial population. The detailed flow chart diagram is shown in Fig. 12.3.

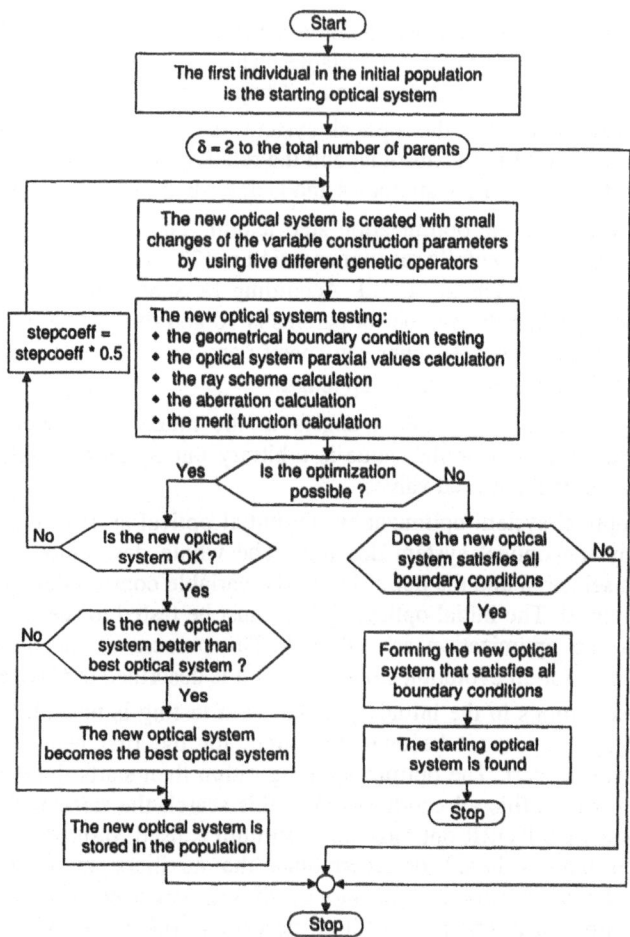

Figure 12.3. The flow chart diagram for the creating and testing of the initial population in the multimembered evolution strategy ES KORR

The first thing that is necessary to do in the creating and testing of the initial population is the copying of the initial optical system into the population. The rest of the population is formed with the optical systems created with small variations of variable construction parameters by applying one of five different genetic operators. The detailed description of creating and testing new values for the variable construction parameters is given in Chapter 10.

After all new values for the variable construction parameters are generated and tested, a new optical system is formed. The new optical system must be tested with the following tests:

– the paraxial value calculation;

– the ray scheme calculation;

– the calculation of aberrations and the merit function.

If a new optical system does pass all the tests it is accepted to the population and its merit function is calculated. The merit function is inherited from the classical damped least squares optimization and it is described in Chapter 8.

In optimization it is very important always to know which optical system is the best meaning which optical system has the minimum aberrations. To be able to know which optical system is the best each new optical system must be compared with the best optical system. The comparison of the optical systems is in fact the comparison of its merit functions. If a new optical system has a smaller merit function than the best optical system then it becomes the best optical system.

If the optimization is not possible meaning that the initial optical system does not fulfill all conditions then a new optical system that fulfills all conditions must be found. In order to find the optical system that fulfills all conditions each new created optical system is tested to see if it fulfills all conditions. If a new optical system fulfills all conditions then the search for the optical system that fulfills all conditions is finished and the optical system becomes the initial optical system for the real optimization in which the aberrations are minimized.

12.4 Creating and testing a new optical system in the multimembered evolution strategy ES KORR

The main part of the optical system optimization by the evolution strategies in general and the multimembered evolution strategy ES KORR in particular is creating and testing an optical system. By creating a new optical systems the optimization algorithm can search the optimization space and find good optical systems that can represent a solution to the problem. The detailed flow chart diagram for creating and testing the optical system in the multimembered evolution strategy ES KORR is shown in Fig. 12.4.

The creating of a new optical system in the multimembered evolution strategy ES KORR is the more complex than in other multimembered evolution strategies (ES GRUP and ES REKO). In order to create a new optical system the following steps are necessary:

– the optical systems, which will represent the parents are randomly chosen from the population. The number of chosen optical systems is:

– two if the selected genetic operator is the discrete recombination of the parent pairs and the intermediary recombination of parent pairs;

– two parents for each variable construction parameter if the selected genetic operator is the discrete recombination of all parents in the population and the intermediary recombination of all parents in the population

– according to the selected genetic operator the optimization step length (the standard deviation) is determined for each variable construction parameter.

– if the rotation of the mutation hyperellipsoid is allowed then, according to the selected genetic operator, the value for the rotation angle of the mutation hyperellipsoid is determined.

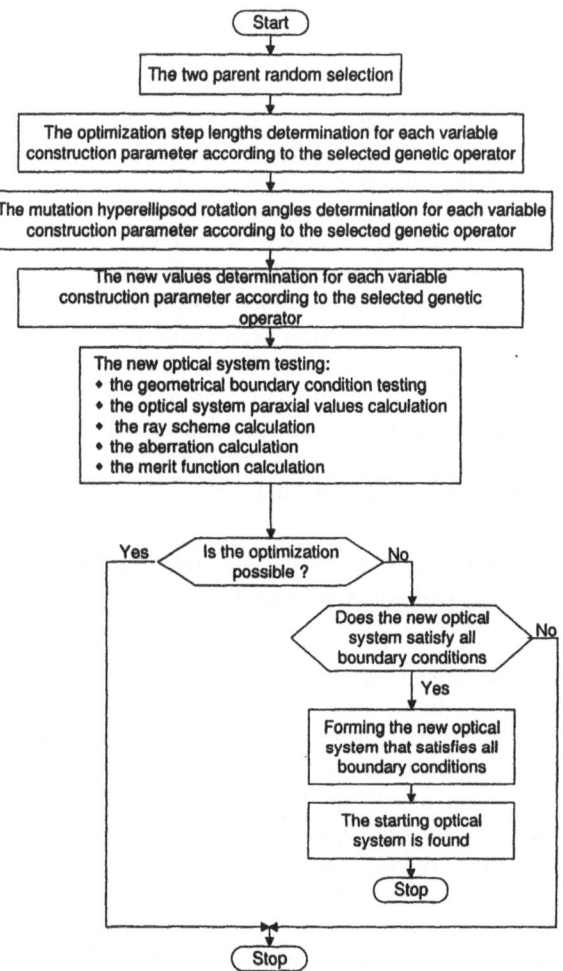

Figure 12.4. The flow chart diagram for the creating and testing a new optical system in the multimembered evolution strategy ES KORR

− all variable optimization parameters (the variable construction parameters, the optimization step lengths and the rotation angles for the mutation hyperellipsod) are randomly changed according to the Gaussian normal distribution law.

− new values for the variable construction parameter are obtained as a sum between the variable part of the construction parameter, which is obtained by the mutation according to the Gaussian normal distribution law, and the fixed part of the construction parameter.

When a new optical system is created it must be tested to be seen whether all conditions are fulfilled. A more detailed description of the optical system testing is given in Chapter 11 and it will not be described here in details.

At the end after the creation and testing of a new optical system the optimization algorithm must decide whether the optimization is possible or not. If the optimization is possible then a new optical system is used in searching the optimization space and finding the optical system with the minimum aberrations. If the optimization is not possible then a new optical system is used in the search for the optical system that fulfills all conditions. This is done by testing each new created optical system to see if it fulfills all conditions. If a new optical system fulfills all conditions then the search for the optical system that fulfills all conditions is finished and that optical system becomes the initial optical system for the real optimization in which the aberrations are minimized.

12.5 Creating a new population in the multimembered evolution strategy ES KORR

After creating a new optical system that satisfies all conditions it is placed in a new population. The process of creating a new population is rather complex because there are two selection methods: the coma strategy and the plus strategy and the number of individuals which is different for the plus and the coma strategy is larger that the number of parents for the next generation. If each individual that satisfies all conditions will be stored in the population then it will be uneconomical use of computer memory. Thus only the best individuals i.e. the optical system with the smallest merit function and aberrations is stored in the population. The number of stored optical systems is equal to the number of the parents for the next generation. The detailed flow chart diagram for creating a new population in the multimembered evolution strategy ES KORR is shown in Fig. 12.5.

The algorithm for creating a new population depends on the selection method for new parents. If the selection method is the coma strategy which means that the parents for the next generation are selected only among the offspring then the algorithm is the same as for the multimembered evolution strategies ES GRUP or ES REKO and it is described in Chapter 11.

If the selection method is the plus strategy which means that the parents for the next generation are selected among the parents for the current generation and their offspring then the algorithm is following:

When a new optical system that satisfies all conditions is created the number of current offspring is increased. There is no direct storing of optical systems in the population because it is not empty but occupied by the parents of the current generation. A new created optical system that satisfies all conditions can be stored in the population if and only if it has the merit function smaller than the biggest merit function from the worst optical system. A new optical system that is stored in the population substitutes the worst optical system. When a new optical system is stored the whole population must be sorted and the worst optical system with the biggest merit function in the population must be found.

When the whole new population is created the best optical system with the smallest merit function is found. This optical system is compared with the best optical system found so far. If the best optical system in a new population has the smaller merit function than the best optical system found so far then the best optical system in a new population becomes the best optical system found so far.

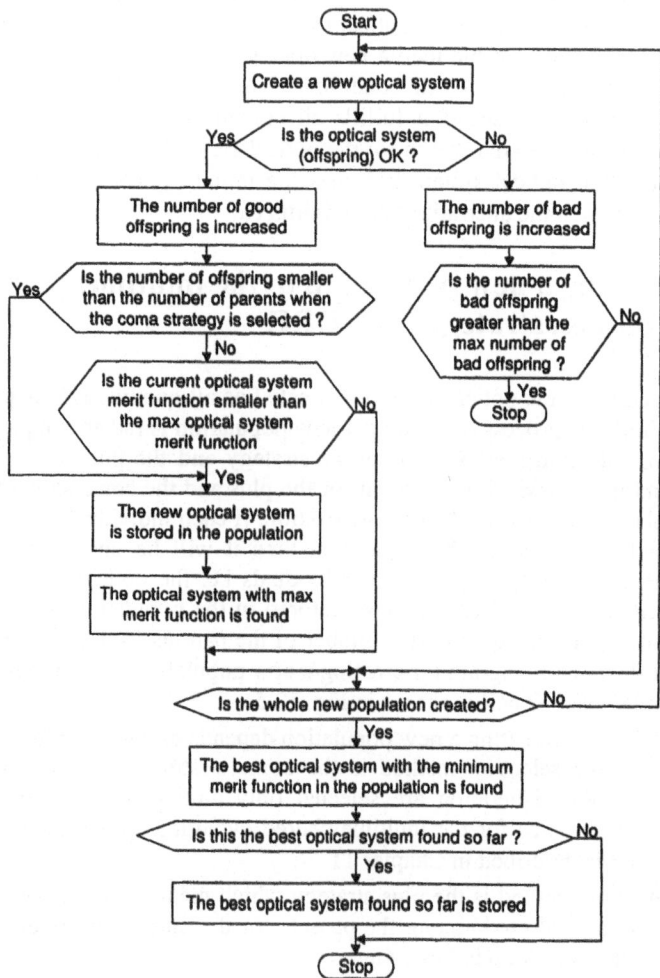

Figure 12.5. The flow chart diagram for creating a new population in the multimembered evolution strategy ES KORR

Chapter 13

The Cooke triplet optimizations

13.1 Introduction

The Cooke triplet consists of three separated lenses positioned at the finite distance. It is often considered that the triplet is one of the most important discoveries in the field of photographic objectives. It is designed by Harold Dennis Taylor in 1893. The Cooke triplet is named after the firm in which H. D. Taylor worked at that time. The Cooke triplet is a very interesting objective for application because it joined the simplicity of construction (it consists of only three lenses) and the possibility of complete aberration correction (it has enough potential variable construction parameters to correct the basic chromatic aberrations and all five primary monochromatic aberrations: the spherical aberration, the coma, the astigmatism, the field curvature and the distortion).

In its original form the Cooke triplet consists of three separated lenses, two outer lenses being biconvex (converging) lenses and between them there is the biconcave (diverging) lens. The aperture diaphragm is usually positioned in the space between the second and the third lens. A high degree of symmetry is present in the Cooke triplet construction. The converging lenses are very similar and they are positioned on the nearly equal distances from the diverging lens. Both refraction surfaces of the diverging lens have nearly equal radii of curvature. The converging lenses are made of crown glass and the diverging lenses are made of flint glass. Usually, the selected glasses are standard glasses that are easy for production.

Standard Cooke triplets have the relative aperture of $f/4$ and the field angle up to maximum $2\omega = 50°$. If the Cooke triplet has relatively larger effective focal length and works with smaller field angles it is possible to design the Cooke triplet with the relative aperture $f/3$. One rule must be obeyed in the Cooke triplet design: the larger relative aperture is desired the smaller field angle can be obtained. The Cooke triplet is usually designed as anastigmatic lens i.e. as the objective with the corrected stigmatism and the field curvature while other primary monochromatic aberrations (the spherical aberration, the coma and the distortion) have sufficiently small values. The astigmatism correction is obtained by appropriate selection of the radii of curvature and the thickness of the converging lenses that are positioned at the finite relatively large distances from the aperture diaphragm. The Petzval sum correction i.e. the field curvature correction is obtained by appropriate selection of the separation between the converging and diverging lenses.

13.2   The Cooke triplet initial design and the optimization goals

The initial design for the Cooke triplet is a standard design taken from the Arthur Cox classical monograph about lens design [46] and it has the label Cox design no. 401. In this monography there is a large number of designed optical systems, mostly patented. These are very good optical systems that are already optimized in order to fulfil requirements for various instruments. All objectives presented in [46] are calculated for the effective focal length of 1". In order to optimize the Cooke triplet in the program ADOS it was necessary to translate all measurement units from the Anglo-Saxon measurement units into the international measurement units. Furthermore, all glasses in the Cooke triplet are selected from the Schott glass catalogue. The Cooke triplet with the international measurement units for all construction data and glasses selected from the Schott glass catalogue is a little different from the original design of the Cooke triplet presented in [46] but that is necessary if the optimizations are wanted to made with the program ADOS. There are two optimization goals for the following optimizations:

- the first optimization goal is obvious: optimize the Cooke triplet in order to obtain the smallest possible minimum of the merit function. There ia a large number of merit function local minima but the goal for the optimization algorithm is to try to find the smallest possible local minimum i.e. the global minimum. The merit function used in all optimizations is defined in Section 8.2.
- the second optimization goal is to compare the optimization designs and to see how many different but good designs of Cooke triplet can be found by optimization algorithms.

All optimization example objectives: the Cooke triplet, the Petzval and the Double Gauss are optimized with the following variable parameters:

- the variable construction parameters;
- the definition of boundary conditions;
- the definition of aberration weights;
- the definition of the paraxial data that is to be controlled and the weighting factor for each controlled paraxial data;
- the variable optimization parameters.

All above stated variable parameters except the variable optimization parameters are connected with the selected optimization example and they are the same for all optimization algorithms. With the classical objectives the possible variable construction parameters are:

- all radii of curvature;
- all radii of curvature and all separations between the optical surfaces;
- all radii of curvature, all separations between the optical surfaces and all glasses.

The definition of boundary conditions consists of:

- the minimum axial separation between optical surfaces;
- the minimum edge separation between optical surfaces;
- the maximum clear aperture for each optical surface.

All variable construction parameters and all definition of boundary conditions are specific for each optimization example. For aberration weights and paraxial data default values are accepted because they give standard objectives optimum values.

Each optimization algorithm has a smaller or a larger set of parameters, which can be varied. Each parameter has influence on the optimization algorithm and the possibility of finding the best local minimum and hopefully the global minimum. The first optimization algorithm is the classical DLS optimization. The most influential parameters are parameters that represent the increments of the variable construction parameters used in the calculation of aberrational partial derivatives with the respect to the construction parameters like:

- the curvature increment;
- the separation increment;
- the glass refraction index and the dispersion increment.

A very influential parameter is also a selection of glasses for the corners of the glass triangle. If during the optimization glasses in optical system are variable then they can move only within the glass triangle.

The ASSGA optimization has the largest set of optimization parameters that can be varied. In this set some parameters have a great influence on the possibility of finding the global minimum and some do not. Here only the most influential parameters will be summarized:

The first and the most influential parameter is the population size. If the population is too small then there are not enough different individuals, which will enable the genetic algorithm to search the optimization space successfully, escaping the local minima and looking for the global minimum. Too big population is not good as well because a larger number of individuals than necessary extends the time needed for the optimization to find the optimum solution. The optimum value for the population size must be found experimentally and in all optimization examples this is done at first place.

The second important parameter is the probabilities of the genetic operators used in the roulette wheel selection. Each genetic operator (the uniform crossover, the average crossover, the floating point number mutation, the big number creep and the small number creep) has its probability and the sum of all probabilities must be always 100. In the Cooke triplet optimization the following sets of genetic operator probabilities are used:

- All genetic operators have equal probability at the start of optimization. Each genetic operator has the probability equal to 20 because there are five genetic operators and the sum of all probabilities is 100. During the optimization the genetic operator probabilities are changed according to the merit function value of the optical system they produced.

– The crossover operators have larger probabilities than the mutation operators at the start of optimization. This means that at the start of optimization, dominant operators in the population are crossover operators. This is a classical genetic algorithms case with the dominant crossover operator and the mutation as a background operator. The uniform crossover operator and the average crossover operator at start of optimization have the probability 30, the floating point number mutation operator at the start of optimization has the probability 20 and the big number creep operator and the small number creep operator at the start of optimization have the probability 10. The minimum probability that each genetic operator can have is also a variable parameter and the default value is 5, which is not changed in any optimization.

– The mutation operators have the larger probabilities than the crossover operators at the start of optimization. This means that the mutation operators are a dominant genetic operators in the population at the start of optimization. This is contrary to the standard genetic algorithms where the crossover is a dominant operator and the mutation is a background operator responsible only for keeping diversity in the population. The floating point number mutation operator at the start of optimization has the probability 30, the big number creep operator and the small number creep operator at the start of optimization have the probability 20, the uniform crossover operator and the average crossover operator at the start of optimization have the probability 15.

All other optimization parameters are not so influential and they are not varied.

The two membered evolution strategy ES EVOL has a rather small set of variable parameters. The default values for all parameters are very well chosen and they are not changed in all optimizations. In the multimembered evolution strategies ES GRUP, ES REKO and ES KORR the default values for all parameters are also used.

In this Chapter all optical systems, initial and optimized are represented only by the optical system merit function value. This is the most convenient representation of the optimization results. For more information about input data files and optimization results reader can contact author on email address darko@afrodita.rcub.bg.ac.yu.

13.3   DLS Optimizations

The mathematical theory of the DLS optimization is given in Chapter 2 and the implementation of this optimization algorithm in the program ADOS is given in Chapter 8. The DLS optimizations for the selected Cooke triplet are calculated for the following conditions:

– all radii of curvature are variable;

– all radii of curvature and all separations between the optical surfaces are variable;

– all radii of curvature, all separations between the optical surfaces and all glasses are variable.

All DLS optimizations are calculated with the following damping algorithms:

– the additive damping;

– the multiplicative damping.

The first DLS optimizations of the Cooke triplet are when all radii of curvature are variable. The Cooke triplet consists of three separated lenses, which makes six possible variable radii of curvature. In the optimization only five radii of curvature were used and the sixth radius of curvature was used for keeping the effective focal length at a predefined value. The merit function of the initial Cooke triplet and the optimized Cooke triplets are given in Table 13.1.

Table 13.1. The merit function values of the optimized Cooke triplets for different damping algorithms. Only radii of curvature are variable.

	Merit function
The initial optical system	4.544
The DLS optimization with the additive damping	4.544
The DLS optimization with the multiplicative damping	4.544

From Table 13.1 it is clear that all Cooke triplet merit functions are equal and that in fact no optimization happened. There are two reasons which explain that situation. The initial optical system is a very good optical system taken from [46]. Only five variable radii of curvature is a too small number of variable parameters and the initial optical system is in fact in one of local minimums so the optimization algorithm does not have the power to escape the local minimum and to search the rest of optimization space. One of possible ways to try to escape this local minimum is to increase the increment with which the variable radii are changed during the calculation of the aberrational partial derivatives with the respect to the construction parameters. The default value for the curvature increment is 0.00001. The idea behind the curvature increment value is to make small changes in the variable radii of curvature to insure linearity. The value for curvature increment is increased to the 0.0001 and 0.001. The merit function values for the optimized Cooke triplets are given in Table 13.2.

Table 13.2. The merit function values of the optimized Cooke triplets for different curvature increments. Only radii of curvature are variable.

	Merit function
The initial optical system	4.544
The DLS optimization with the curvature increment 0.00001	4.544
The DLS optimization with the curvature increment 0.0001	4.544
The DLS optimization with the curvature increment 0.001	4.546

From Table 13.2 it can be seen that for sufficiently small values of the curvature increment (0.00001 and 0.0001) the linearity is conserved and the optimization algorithm finds the same local minimum. When the curvature increment is large enough (0.001), the supposed linearity is not conserved and the optimization algorithm cannot find the same local minimum. Instead, it finds the local minimum with the greater merit function.

The next series of DLS optimizations are done when all radii of curvature and all separations between the optical surfaces are variable. Now the optimization algorithm has a possibility to change 11 construction parameters (5 radii of curvature and 6 separations between the optical surfaces). The merit functions of the initial optical system and the optimized optical systems are given in Table 13.3.

Table 13.3. The merit function values of the optimized Cooke triplet for different damping algorithms. The radii of curvature and the separations between optical surfaces are variable.

	Merit function
The initial optical system	4.544
The DLS optimization with the additive damping	0.566
The DLS optimization with the multiplicative damping	0.482

When all radii of curvature and separations between the optical surfaces are variable, the optimization algorithm has enough variable construction parameters to search the optimization space for a better local minimum. From Table 13.3 it can be seen that the DLS optimization with the multiplicative damping has found a slightly better optical system with the merit function 0.482.

The last series of DLS optimizations are done when all construction parameters are variable (all radii of curvature, all separations between the optical surfaces and all glasses). The total number of variable construction parameters is 14. The DLS optimization algorithm expects that the variable parameters are continuous parameters. The radii of curvature and the separations between the optical surfaces are continuous parameters but the glasses are discrete parameters (i.e. there is a finite number of glasses in the manufactures glass database). The variable parameter ought to be continuous because the partial derivatives of the aberrations with the respect to the variable construction parameters are calculated by adding and subtracting increments from the variable parameter and calculating aberrations. The discrete parameter must be transformed to the continuous parameter and the whole optimization must be done with this continuous parameter. The DLS optimization algorithm usually finds a good local minimum with this continuous parameter, but when these continuous parameters are transformed to the real glasses from the glass database, the merit function of this real optical system may be different from the merit function of the optical system with the variable glasses as continuous parameters. The conversion from the glass continuous parameter to the nearest real glass is done so that:

– the difference between the calculated index of refraction and the real glass index of refraction is minimum;

– the difference between the calculated Abbe number and the real glass Abbe number is minimum.

In the author's opinion the DLS optimization with the variable glasses ought to be used only when it is absolutely necessary. The optical designer can select good glasses from the glass database and run the DLS optimization with only the radii of curvature and separations between the optical surfaces variable.

The merit functions of the initial optical system and the optimized optical systems are given in Table 13.4.

Table 13.4. The merit function values of the optimized Cooke triplet for different damping algorithms. The radii of curvature, the separations between optical surfaces and the glasses are variable.

	Merit function
The initial optical system	4.544
The DLS optimization with the additive damping	0.963
The DLS optimization with the multiplicative damping	0.526

In Table 13.4 the merit function is given for the glasses as continuous parameters. The DLS optimization with the multiplicative damping is managed to find an optical system with a better merit function than the DLS optimization with the additive damping. This optical system found by the DLS optimization with the multiplicative damping has also a better merit function when the glasses as continuous parameters are changed with the real glasses.

13.4   ASSGA optimizations

The mathematical theory of the ASSGA optimization is given in Chapter 3 and the implementation of this optimization algorithm in the program ADOS is given in Chapter 9. The ASSGA optimizations for the selected Cooke triplet are calculated for the following conditions:

— all radii of curvature are variable;
— all radii of curvature and all separations between the optical surfaces are variable;
— all radii of curvature, all separations between the optical surfaces and all glasses are variable.

In all ASSGA optimizations the following strategy is accepted. Each series of the ASSGA optimizations is started with all default parameters. The first optimization parameter that ought to be varied is the population size because there is no generally accepted default value for the population size.

Since the probabilistic events play an important role in the ASSGA optimization, it is necessary to run several optimizations to see a general behaviour of the optimization algorithm. The author decided to run five optimizations for each changed optimization parameter.

The first ASSGA optimizations of the Cooke triplet are calculated when all radii of curvature are variable. There are only five radii of curvature that can be varied. As in the DLS optimizations this is a rather small number of variable construction parameters and the ASSGA optimization algorithm cannot be expected to find optical systems with significantly better merit function than the initial optical system merit function. The merit functions of optimized optical systems for different population sizes are given in Table 13.5. All genetic operator probabilities have identical values equal to 20.

Table 13.5. The merit functions of the optimized Cooke triplets for different population sizes. Only radii of curvature are variable.

	population 200	population 300	population 400
The 1st optimization	4.510	4.727	4.489
The 2nd optimization	4.538	4.532	4.522
The 3rd optimization	4.650	4.481	4.526
The 4th optimization	4.528	4.507	5.313
The 5th optimization	4.572	4.534	4.511
The merit function mean value	4.560	4.556	4.672
The merit function standard deviation	0.055	0.098	0.358

From Table 13.5 it can be seen that the ASSGA optimization has found optical systems with similar merit functions. If the merit function mean value is a criterion for the population size selection then the population size of 300 optical systems has the smallest merit function mean value.

After the population size is determined, the next step is the variation of the starting values for the genetic operators probabilities. Three sets of the genetic operator probabilities are discussed.

- The first set is the default values for all genetic operator probabilities. The starting values for the genetic operator probabilities are the following: the average crossover, the uniform crossover, the floating point number mutation, the big number creep and the small number creep are equal to 20 ($AC = UC = FM = BM = SM = 20$).

- The second set is the set with larger values for the crossover probabilities than for the mutation probabilities. The starting values for the genetic operator probabilities are the following: the average crossover and the uniform crossover have the probability 30, the floating point number mutation has the probability 20, the big number creep and the small number creep have the probability 10 ($AC = UC = 30$, $FM = 20$, $BM = SM = 10$).

- The third set is the set with larger values for the mutation probabilities than for the crossover probabilities. The starting values for the genetic operator probabilities are the following: the average crossover and the uniform crossover have the probability 15, the floating point number mutation has the probability 30, the big number creep and the small number creep have the probability 20 ($AC = UC = 30$, $FM = 20$, $BM = SM = 10$).

The merit functions of the optimized optical systems for different starting values for the genetic operators probabilities are given in Table 13.6. The population size is 300 individuals (optical systems).

Table 13.6. The merit functions of the optimized Cooke triplets for different starting values of the genetic operators probabilities. Only radii of curvature are variable.

	1st opt.	2nd opt.	3rd opt.	4th opt.	5th opt.	mean value	stand. deviat.
$AC = UC = FM = BM = SM = 20$	4.727	4.532	4.481	4.507	4.534	4.556	0.098
$AC = UC = 30$, $FM = 20$, $BM = SM = 10$	4.513	4.809	4.871	4.472	4.523	4.638	0.187
$AC = UC = 15$, $FM = 30$, $BM = SM = 20$	4.598	4.534	4.623	4.515	5.011	4.656	0.203

where:

AC is the average crossover;

UC is the uniform crossover;

FM is the floating point number mutation;

BM is the big number creep;

SM is the small number creep.

From Table 13.6 it can be seen that the ASSGA optimization has found optical systems with similar merit functions. When the starting values for all genetic operator probabilities are equal then the ASSGA optimization finds optical systems having the smallest merit function mean value and the standard deviation. This means that all found optical systems are grouped around one local minimum. When the starting values for the crossover operator probabilities are larger than for the mutation probabilities, then an optical system with the smallest merit function is found. When the crossover operators and the mutation operators have different values for the starting probabilities the merit function mean value and the standard deviation is larger than the corresponding values when all genetic operator probabilities have equal values.

The second set of the Cooke triplet ASSGA optimizations is calculated when all radii of curvature and all separations between the optical surfaces are variable. The number of variable construction parameters is 11 (5 variable radii of curvature and 6 separations between the optical surfaces). The ASSGA optimization algorithm has enough variable construction parameters to search successfully the optimization space and find optical systems with a significantly smaller merit function than the initial optical system merit function. The first step, as in all ASSGA optimizations, is the determination of the population size. All other optimization parameters have default values. The merit functions of optimized optical systems for different population sizes are given in Table 13.7.

Table 13.7. The merit functions of the optimized Cooke triplets for different population sizes. The radii of curvature and the separations between the optical surfaces are variable.

	population 400	population 500	population 600
The 1st optimization	0.894	0.578	0.839
The 2nd optimization	0.664	0.667	1.190
The 3rd optimization	1.207	0.790	0.746
The 4th optimization	1.073	0.960	0.772
The 5th optimization	0.769	0.761	0.726

	population 400	population 500	population 600
The merit function mean value	0.921	0.751	0.855
The merit function standard deviation	0.221	0.143	0.192

From Table 13.7 it can be seen that the ASSGA optimization algorithm has found optical systems with quite different merit functions that differ from 0.578 to 1.207 for different population sizes. The optimum population size that has the minimum merit function mean value and the standard deviation is 500 optical systems.

After the optimum population size is determined the next step is variation of the starting values for the genetic operators probabilities. The sets of starting values for the genetic operator probabilities are the same as in case when the variable are only radii of curvature. The merit functions of optimized optical systems for different starting values of the genetic operators probabilities are given in Table 13.8. The population size is 500 individuals (optical systems).

Table 13.8. The merit functions of the optimized Cooke triplets for different starting values of the genetic operators probabilities. The radii of curvature and the separations between the optical surfaces are variable.

	1st opt.	2nd opt.	3rd opt.	4th opt.	5th opt.	mean value	stand. deviat.
$AC = UC = FM = BM = SM = 20$	0.578	0.667	0.790	0.960	0.761	0.751	0.143
$AC = UC = 30$, $FM = 20$, $BM = SM = 10$	1.043	1.096	0.814	1.198	1.149	1.060	0.149
$AC = UC = 15$, $FM = 30$, $BM = SM = 20$	0.773	0.786	0.612	0.871	0.946	0.798	0.125

where:

AC is the average crossover;

UC is the uniform crossover;

FM is the floating point number mutation;

BM is the big number creep;

SM is the small number creep.

From Table 13.8 it can be seen that the ASSGA optimization algorithm has found optical systems with similar merit functions when the starting values for the genetic operator probabilities are equal and when the mutation operators are larger than the crossover operators. The optical systems found by the ASSGA optimization algorithm when the crossover operators are lager than the mutation operators have significantly larger merit functions. An optical system with the minimum merit function (0.578) is found by the ASSGA optimization when the starting values for the genetic operator probabilities are equal to 20.

The third set of the ASSGA optimizations of the Cooke triplet is calculated when all radii of curvature, all separations between the optical surfaces and all glasses are variable. The number of variable construction parameters is 14 (5 variable radii of curvature, 6 separations between the optical surfaces and 3 glasses). The ASSGA optimization algorithm does not require that all variables be continuous, so the ASSGA optical system can work directly with glasses from the glass database. In fact the ASSGA optimization algorithm randomly selects a glass from the glass database. The ASSGA optimization algorithm has enough variable construction parameters to search the optimization space successfully and find optical systems with the best merit function. The ASSGA optimization has found much better optical systems with a smaller merit function when the glasses are variable than when the glasses are constant (not variable). Other optimization algorithms usually have found better optical systems when only the radii of curvature and the separations between the optical surfaces are variable. The first step, as in all ASSGA optimizations, is the population size determination. All other optimization parameters have default values. The merit functions of optimized optical systems for different population sizes are given in Table 13.9.

Table 13.9. The merit functions of the optimized Cooke triplets for different population sizes. The radii of curvature, the separations between the optical surfaces and the glasses are variable.

	populat. 500	populat. 600	populat. 700	populat. 800
The 1st optimization	1.275	0.792	1.350	1.007
The 2nd optimization	1.387	0.683	0.792	0.398
The 3rd optimization	0.962	1.136	0.532	1.114
The 4th optimization	0.930	0.861	0.609	0.491
The 5th optimization	0.808	1.185	0.580	1.193
The merit function mean value	1.078	0.931	0.773	0.841
The merit function standard deviation	0.246	0.219	0.337	0.369

From Table 13.9 it can be seen that the ASSGA optimization algorithm has found optical systems with the quite different merit functions from 0.398 to 1.387. There is no concentration around one local minimum like in preceding optimizations. In the ASSGA optimization more variable construction parameters mean more chance to find a good local minimum. With all 14 construction parameters variable the ASSGA optimization has found the optical system with the minimum merit function which is 0.398. The optimum population size that has the minimum merit function mean value and the standard deviation is 700 optical systems. The population size that allowed the ASSGA optimization to find optical systems with the smallest merit functions (0.398 and 0.491) is 800 optical systems.

After the optimum population size is determined the next step is the variation of the starting values for the genetic operators probabilities. The sets of starting values for the genetic operator probabilities are the same in all optimizations when the variables are only radii of curvature and when the variable are the radii of curvature and the separations between the optical surfaces. The merit functions of optimized optical systems for different starting values for the genetic operators probabilities are given in Table 13.10. The population size is 700 optical systems.

Table 13.10. The merit functions of the optimized Cooke triplets for different starting values of the genetic operators probabilities. The radii of curvature, the separations between the optical surfaces and the glasses are variable.

	1st opt.	2nd opt.	3rd opt.	4th opt.	5th opt.	mean value	stand. deviat.
$AC = UC = FM = BM = SM = 20$	1.350	0.792	0.532	0.609	0.580	0.773	0.337
$AC = UC = 30, FM = 20, BM = SM = 10$	1.682	1.488	0.728	1.313	0.679	1.178	0.453
$AC = UC = 15, FM = 30, BM = SM = 20$	1.006	0.864	1.602	0.936	0.566	0.995	0.379

where:

AC is the average crossover;

UC is the uniform crossover;

FM is the floating-point number mutation;

BM is the big number creep;

SM is the small number creep.

From Table 13.10 it can be seen that the ASSGA optimization has found optical systems with different merit functions. When the starting values for genetic operator probabilities are identical and equal to the 20 the ASSGA optimization has found optical systems with the smallest merit function so it can be concluded that the optimum value for the genetic operator probabilities is to be identical and equal to 20. For all other combinations of genetic operator probabilities the ASSGA optimization has found the optical systems that have larger merit functions.

13.5   Evolution strategies optimizations

The evolution strategies are global optimization algorithms based on the theory of evolution and the natural selection developed for the optimization of complex technical systems. In this monograph the mathematical theory for four different evolution strategies is given in Chapter 4 and their implementation in the program ADOS is given in Chapters 10 – 12. Here will be presented the optimizations of the selected Cooke triplet with the following evolution strategies:

– the two membered evolution strategy ES EVOL;

– the multimembered evolution strategy ES GRUP;

– the multimembered evolution strategy ES REKO;

– the multimembered evolution strategy ES KORR.

13.5.1   Two membered evolution strategy ES EVOL optimizations

The two-membered evolution strategy ES EVOL represents the most simple evolution strategy. The ES EVOL optimizations for the selected Cooke triplet are calculated for the following conditions:

– all radii of curvature are variable;

- all radii of curvature and all separations between the optical surfaces are variable;
- all radii of curvature, all separations between the optical surfaces and all glasses are variable.

The two membered evolution strategy ES EVOL has a small number of optimization parameters and all parameters are kept at default values.

The first ES EVOL optimizations of the Cooke triplet are calculated when all radii of curvature are variable. There are only five radii of curvature that can be varied. As in the DLS and the ASSGA optimizations this is a rather small number of variable construction parameters and the ES EVOL optimization algorithm cannot be expected to find optical systems with significantly better merit function than the initial optical system merit function. The merit functions of optimized optical systems are given in Table 13.11.

It is very interesting that, although the probability plays an important part in the ES EVOL optimization, all five optimizations have found the same optical system with the merit function 4.462. This can be explained that the optical system is a good optical system representing the local minimum in the optimization space. The optimization algorithm does not have enough variable construction parameters to escape from this local minimum and to search the rest of the optimization space.

Table 13.11. The merit functions of the optimized Cooke triplets. Only radii of curvature are variable.

	Merit function
The 1st optimization	4.462
The 2nd optimization	4.462
The 3rd optimization	4.462
The 4th optimization	4.462
The 5th optimization	4.462
The merit function mean value	4.462
The merit function standard deviation	0.000

The second group of the ES EVOL optimizations of the Cooke triplet is calculated when all radii of curvature and all separations between the optical surfaces are variable. The number of variable construction parameters is 11 (5 variable radii of curvature and 6 separations between the optical surfaces). The ES EVOL optimization algorithm has enough variable construction parameters to search the optimization space successfully and find optical systems with a significantly smaller merit function than the initial optical system merit function. The merit functions of optimized optical systems are given in Table 13.12.

From Table 13.12 it can be seen that all ES EVOL optimisations have found optical systems with very similar merit functions. The merit function mean value is 0.476 and the standard deviation is only 0.008.

Table 13.12. The merit functions of the optimized Cooke triplets. The radii of curvature and the separations between the optical surfaces are variable.

	Merit function
The 1st optimization	0.473
The 2nd optimization	0.468
The 3rd optimization	0.489
The 4th optimization	0.476
The 5th optimization	0.475
The merit function mean value	0.476
The merit function standard deviation	0.008

The third group of ES EVOL optimizations of the Cooke triplet is calculated when all radii of curvature, all separations between the optical surfaces and all glasses are variable. The number of variable construction parameters is 14 (5 variable radii of curvature, 6 separations between the optical surfaces and 3 glasses). The evolution strategies are originally designed for the continuous variable optimization. The problem rises because the variable glasses are not continuous parameters. They are discrete parameters that are selected from the database with the finite number of different glasses. In the optimization the glass variation (mutation) consists of two steps. In the first step glasses are treated as all other variable construction parameters. The glass index of refraction and the Abbe number are mutated as other continuous parameters. In the second step the glass database is searched to find the corresponding glass to the new obtained values for the glass index of refraction and the Abbe number. The new glass must be of the same type meaning if the starting glass is the crown glass then the new changed (mutated) glass must be also the crown glass. If the mutated glass is not the same type as the starting glass then the mutation is rejected and another one is calculated. This method of variable glass mutation gives a large number of rejected mutations and the 1/5 rule is not any more fulfilled. The 1/5 rule states that in average one of five mutations ought to be successful meaning it ought to produce an optical system that has a better merit function than the starting optical system. The consequence of not fulfilling the 1/5 rule is very slow convergence to the minimum which is not the best possible local minimum and the large number of executed mutations with the small number of successful mutations. The merit functions of the optimized optical systems are given in Table 13.13.

Table 13.13. The merit functions of the optimized Cooke triplets. The radii of curvature, the separations between the optical surfaces and the glasses are variable.

	Merit function
The 1st optimization	0.798
The 2nd optimization	0.707
The 3rd optimization	0.721
The 4th optimization	0.645
The 5th optimization	0.751

	Merit function
The merit function mean value	0.724
The merit function standard deviation	0.056

From Table 13.13 it can be seen that all ES EVOL optimizations have found optical systems with the similar merit functions. The merit function standard deviation is only 0.056. It is interesting to note that although there are more variable construction parameters, the ES EVOL optimization algorithm was not able to find optical systems with a better merit function than when the variable were only the radii of curvature and the separations between the optical surfaces. In fact the ES EVOL optimization algorithm has found optical systems with more than 50% worse merit functions. The merit function mean value for the optical systems with variable radii of curvature and the separations between the optical surfaces is 0.476 and the merit function mean value for the optical systems with variable radii of curvature, the separations between the optical surfaces and glasses is 0.724.

13.5.2 Multimembered evolution strategy ES GRUP optimizations

The multimembered evolution strategy ES GRUP represents the further development of ideas from the two membered evolution strategy ES EVOL. The ES GRUP optimizations for the selected Cooke triplet are calculated for the following conditions:

− all radii of curvature are variable;
− all radii of curvature and all separations between the optical surfaces are variable;
− all radii of curvature, all separations between the optical surfaces and all glasses are variable.

The multimembered evolution strategy ES GRUP has a small number of optimization parameters and all parameters are kept at default values.

The first ES GRUP optimizations of the Cooke triplet are calculated when all radii of curvature are variable. There are only five radii of curvature that can be varied. As in all preceding optimizations this is a rather small number of variable construction parameters and the ES GRUP optimization algorithm cannot be expected to find optical systems with a significantly better merit function than the initial optical system merit function. The merit functions of the optimized optical systems are given in Table 13.14.

Table 13.14. The merit functions of the optimized Cooke triplets. Only radii of curvature are variable.

	Merit function
The 1st optimization	4.465
The 2nd optimization	4.463
The 3rd optimization	4.468

	Merit function
The 4th optimization	4.462
The 5th optimization	4.464
The merit function mean value	4.464
The merit function standard deviation	0.002

All ES GRUP optimizations have found almost identical optical systems with the merit functions form 4.462 to 4.468 (the mean value is 4.464 and the standard deviation is 0.002). The explanation of these results is same as given for the ES EVOL optimizations. The multimembered evolution strategy ES GRUP optimization has a too small number of variable construction parameters to escape from the local minimum and to search the optimization space successfully.

The second group of ES GRUP optimizations of the Cooke triplet is calculated when all radii of curvature and all separations between the optical surfaces are variable. The number of variable construction parameters is 11 (5 variable radii of curvature and 6 separations between the optical surfaces). Now the ES GRUP optimization algorithm has enough variable construction parameters to search the optimization space successfully and find optical systems with a significantly smaller merit function than the initial optical system merit function. The merit functions of the optimized optical systems are given in Table 13.15.

Table 13.15. The merit functions of the optimized Cooke triplets. The radii of curvature and the separations between the optical surfaces are variable.

	Merit function
The 1st optimization	0.534
The 2nd optimization	0.501
The 3rd optimization	0.502
The 4th optimization	0.496
The 5th optimization	0.522
The merit function mean value	0.511
The merit function standard deviation	0.016

The ES GRUP has found an optical system with very similar merit functions. It is interesting to note that the ES GRUP optimization algorithm which represents further development of the ES EVOL optimization algorithm, has found optical systems with worse merit functions than the ES EVOL. The ES EVOL optimization algorithm used the population of only two individuals (the parent and the offspring) and the ES GRUP optimization algorithm used the population of 10 parents and 100 offspring. Both optimization algorithms, the ES EVOL and the ES GRUP, used only one genetic operator – the mutation. The merit function mean value for the optical systems found by the ES EVOL optimization is 0.476 and the merit function mean value for the optical systems found by the ES GRUP optimization is 0.511.

The third group of the ES GRUP optimizations of the Cooke triplet is calculated when all radii of curvature, all separations between the optical surfaces and all glasses are variable. The number of variable construction parameters is 14 (5 variable radii of curvature, 6 separations between the optical surfaces and 3 glasses). The problem with the variable glasses is the same as in the ES EVOL optimization described in Section 13.5.1. The consequence of this problem is a very slow convergence to the minimum which is not the best possible local minimum and a large number of executed mutations with the small number of successful mutations. The merit functions of the optimized optical systems are given in Table 13.16.

Table 13.16. The merit functions of the optimized Cooke triplets. The radii of curvature, the separations between the optical surfaces and the glasses are variable.

	Merit function
The 1st optimization	0.759
The 2nd optimization	0.848
The 3rd optimization	0.952
The 4th optimization	0.837
The 5th optimization	1.047
The merit function mean value	0.889
The merit function standard deviation	0.112

The ES GRUP optimizations have found optical systems with the different merit functions from 0.759 to 1.047. Since the variable glasses are not continuous parameters, the ES GRUP optimizations have found optical systems with worse merit functions than the ES GRUP optimizations when are variable only the radii of curvature and the separations between the optical surfaces. The merit function mean value for the optical systems with variable radii of curvature and the separations between the optical surfaces is 0.511 and the merit function mean value for the optical systems with variable radii of curvature, the separations between the optical surfaces and glasses is 0.889. It is interesting to note that the multimembered evolution strategy ES GRUP has found optical systems with a worse merit functions than the two membered evolution strategy ES EVOL. The corresponding merit function mean values for optical systems are 0.724 for the ES EVOL and 0.889 for the ES GRUP.

13.5.3 Multimembered evolution strategy ES REKO optimizations

The multimembered evolution strategy ES REKO represents the further development of ideas from the two membered evolution strategy ES EVOL and the multimebered evolution strategy ES GRUP. The ES REKO optimizations for the selected Cooke triplet are calculated for the following conditions:

- all radii of curvature are variable;
- all radii of curvature and all separations between the optical surfaces are variable;
- all radii of curvature, all separations between the optical surfaces and all glasses are variable.

The multimembered evolution strategy ES REKO has the same set of optimization parameters as the ES GRUP optimization and all parameters are kept at default values.

The first ES REKO optimizations of the Cooke triplet are calculated when all radii of curvature are variable. There are only five radii of curvature that can be varied. As in all preceding optimizations this is a rather small number of variable construction parameters and the ES REKO optimization algorithm can not be expected to find optical systems with significantly better merit function than the initial optical system merit function. The merit functions of optimized optical systems are given in Table 13.17.

Table 13.17. The merit functions of the optimized Cooke triplets. Only radii of curvature are variable.

	Merit function
The 1st optimization	4.466
The 2nd optimization	4.462
The 3rd optimization	4.462
The 4th optimization	4.464
The 5th optimization	4.462
The merit function mean value	4.463
The merit function standard deviation	0.002

All ES REKO optimizations have found almost identical optical systems with the merit functions form 4.462 to 4.466 (the mean value is 4.463 and the standard deviation is 0.002). The explanation of these results is same as given for preceding evolution strategies (the ES EVOL and the ES GRUP) optimizations. The ES REKO optimization has a too small number of variable construction parameters to escape from local minimum and to search the optimization space successfully.

The second group of ES REKO optimizations of the Cooke triplet is calculated when all radii of curvature and all separations between the optical surfaces are variable. The number of variable construction parameters is 11 (5 variable radii of curvature and 6 separations between the optical surfaces). Now the ES REKO optimization algorithm has enough variable construction parameters to search the optimization space successfully and find optical systems with a significantly smaller merit function than the initial optical system merit function. The merit functions of the optimized optical systems are given in Table 13.18.

All ES REKO optimizations of the Cooke triplet have found almost identical optical systems with very similar merit function values. The merit function values range from 0.499 to 0.513 with the mean value 0.508 and the standard deviation 0.006. This is almost equal to the ES GRUP optimizations with the merit function mean value 0.511 but worse than the ES EVOL optimizations with the merit function mean value 0.476. It is interesting that the simplest variant of the evolution strategies has found optical systems with the best merit functions.

Table 13.18. The merit functions of the optimized Cooke triplets. The radii of curvature and the separations between the optical surfaces are variable.

	Merit function
The 1ˢᵗ optimization	0.510
The 2ⁿᵈ optimization	0.506
The 3ʳᵈ optimization	0.513
The 4ᵗʰ optimization	0.512
The 5ᵗʰ optimization	0.499
The merit function mean value	0.508
The merit function standard deviation	0.006

The third group of ES REKO optimizations of the Cooke triplet is calculated when all radii of curvature, all separations between the optical surfaces and all glasses are variable. The number of variable construction parameters is 14 (5 variable radii of curvature, 6 separations between the optical surfaces and 3 glasses). The problem with the variable glasses is the same as in all evolution strategies and it is described in Section 13.5.1. The consequence of this problem is a very slow convergence to the minimum which is not the best possible local minimum. The other characteristic of these optimizations is a large number of executed genetic operators (the mutation and the recombination) with a small number of successful genetic operators. The merit functions of the optimized optical systems are given in Table 13.19.

Table 13.19. The merit functions of the optimized Cooke triplets. The radii of curvature, the separations between the optical surfaces and the glasses are variable.

	Merit function
The 1ˢᵗ optimization	0.592
The 2ⁿᵈ optimization	0.720
The 3ʳᵈ optimization	0.627
The 4ᵗʰ optimization	0.685
The 5ᵗʰ optimization	0.589
The merit function mean value	0.642
The merit function standard deviation	0.058

From Table 13.19 it can be seen that the ES REKO optimizations have found optical systems with the merit function values from 0.589 to 0.720. The merit function mean value is 0.642 and the standard deviation is 0.058. The bad influence of the variable glasses, which are not continuous parameters, is the smallest in the ES REKO optimizations. The ES REKO optimizations have managed to find the optical systems with the best merit function values comparing to the ES EVOL and the ES GRUP.

13.5.4   Multimembered evolution strategy ES KORR optimizations

The multimembered evolution strategy ES KORR is the most developed and the most complex evolution strategy. The ES KORR optimizations for the selected Cooke triplet are calculated for the following conditions:
- all radii of curvature are variable;
- all radii of curvature and all separations between the optical surfaces are variable;
- all radii of curvature, all separations between the optical surfaces and all glasses are variable.

All optimization parameters in the multimembered evolution strategy ES KORR are kept at default values.

The first ES KORR optimizations of the Cooke triplet are calculated when all radii of curvature are variable. There are only five radii of curvature that can be varied. As in all preceding optimizations this is a rather small number of variable construction parameters and the ES KORR optimization algorithm can not be expected to find optical systems with a significantly better merit function than the initial optical system merit function. The merit functions of the optimized optical systems are given in Table 13.20.

Table 13.20. The merit functions of the optimized Cooke triplets. Only radii of curvature are variable.

	Merit function
The 1st optimization	4.463
The 2nd optimization	4.463
The 3rd optimization	4.463
The 4th optimization	4.462
The 5th optimization	4.463
The merit function mean value	4.463
The merit function standard deviation	0.000

It is very interesting that, although the probability plays an important part in the ES KORR optimization, four out of five optimizations have found the same optical system with the merit function 4.463. The explanation of these results is the same as given for the two membered evolution strategy ES EVOL and all previous multimembered evolution strategies. The ES KORR optimization has a too small number of variable construction parameters to escape from the local minimum and to search the optimization space successfully.

The second group of ES KORR optimizations of the Cooke triplet is calculated when all radii of curvature and all separations between the optical surfaces are variable. The number of variable construction parameters is 11 (5 variable radii of curvature and 6 separations between the optical surfaces). Now the ES KORR optimization algorithm has enough variable construction parameters to search the optimization space successfully and find optical systems with a significantly smaller merit function than the initial optical system merit function. The merit functions of the optimized optical systems are given in Table 13.21.

Table 13.21. The functions of the optimized Cooke triplets. The radii of curvature and the separations between the optical surfaces are variable.

	Merit function
The 1st optimization	0.497
The 2nd optimization	0.536
The 3rd optimization	0.557
The 4th optimization	0.468
The 5th optimization	0.526
The merit function mean value	0.517
The merit function standard deviation	0.035

From Table 13.21 it can be seen that the ES KORR optimizations have found the optical systems with the merit function values from 0.468 to 0.557. The merit function mean value is 0.517, which is similar to the merit function mean values found by the ES GRUP and the ES REKO optimizations. It is interesting to note that the most complex evolution strategy ES KORR and the simplest evolution strategy ES EVOL have found the optical system with the minimum merit function (0.468).

The third group of ES KORR optimizations of the Cooke triplet is calculated when all radii of curvature, all separations between the optical surfaces and all glasses are variable. The number of variable construction parameters is 14 (5 variable radii of curvature, 6 separations between the optical surfaces and 3 glasses).

The ES KORR optimization has the possibility to rotate the mutation hyperellipsoid. The mutation hyperellipsoid rotation in fact means that a special procedure performs coordinate transformation of the variable construction parameters modification vector. The modification vector contains the mutation increments that are added to the variable construction parameters in order to obtain new values for the variable construction parameters. Initially the components of the modification vector are mutually independent, but they become linearly related after the series of rotations specified by the positional angles. The coordinate transformation involves n_p partial rotations in each of which only two components of the modification vector are involved. According to Schwefel in [21] this feature tries to correct the slow convergence of the two membered evolution strategy in some special cases. In the optical system optimization this feature worked fine when the variable were the radii of curvature and the separations between the optical surfaces because they are continuous variables having a large range of possible values. When the glasses are added to the variable parameters a problem appears because the glasses are discrete variables with a very narrow range of possible values. In order to make the ES KORR optimization possible, one value of the mutation increment must be applied to the variable radii of curvature and the separations between the optical surfaces and other much smaller value of the mutation increment must be applied to the variable glasses. If the mutation hyperellipsoid rotation feature is turned on, then the coordinate transformation mixes mutation increments and the variable glasses get the mutation increments with the larger values and the radii of curvature and the separations between the optical surfaces get the mutation increments with the smaller values. This is not what is desired. In some cases (the optimization examples) this ill calculated mutation increments will allow the ES KORR optimization to continue and in other the optimization will not be possible. Even if the optimization can continue it will converge slowly to the local minimum that is not the best possible local minimum.

In the case of the Cooke triplet the ES KORR optimizations are calculated for both instances: when the mutation hyperellipsoid can rotate and cannot rotate. The merit functions of the optimized optical systems are given in Table 13.22.

Table 13.22. The merit functions of the optimized Cooke triplets when the mutation hyperellipsoid can rotate and cannot rotate. The radii of curvature, the separations between the optical surfaces and the glasses are variable.

	The mutation hyperellipsoid can rotate	The mutation hyperellipsoid cannot rotate
The 1st optimization	0.817	1.057
The 2nd optimization	0.703	0.413
The 3rd optimization	0.654	0.839
The 4th optimization	0.753	0.703
The 5th optimization	0.800	0.920
The merit function mean value	0.745	0.786
The merit function standard deviation	0.068	0.245

From the optimization results presented in Table 13.22 it is obvious that the ES KORR optimizations when the mutation hyperellipsoid can rotate have found optical systems with the similar merit functions that are not the best local minimum. The found optical systems have a worse merit function than the optical systems found by the ES KORR optimizations when the variable were all radii of curvature and all separations between the optical surfaces. When the mutation hyperellipsoid cannot rotate the ES KORR optimizations have found optical systems with the significantly different merit functions from 0.413 to 1.057.

It is interesting to note that only in this case when the mutation hyperellipsoid cannot rotate one evolution strategy with variable glasses has found an optical system with a smaller merit function than the evolution strategy without variable glasses i.e. with only variable the radii of curvature and the separations between the optical surfaces. The optical system with the merit function 0.413 is the best optical system found by any evolution strategy.

13.6   Comparison of the Cooke triplet designs by various optimization methods

All optimization methods: the DLS optimization, the ASSGA optimization, the two membered evolution strategy ES EVOL optimization, the multimembered evolution strategies ES GRUP, ES REKO and ES KORR optimizations have calculated various Cooke triplets with the following variable construction parameters:

– all radii of curvature;
– all radii of curvature and all separations between the optical surfaces;
– all radii of curvature and all separations between the optical surfaces and all glasses.

The first group of the Cooke triplet optimizations is calculated when all radii of curvature are variable. There are only five radii of curvature that can be varied. This is a rather small number of variable construction parameters and any optimization algorithm cannot be expected to find optical systems with a significantly better merit function than the initial optical system merit function. The optical system with the best merit function found by any optimization algorithm, the merit function mean value and the merit function standard deviation calculated by any optimization algorithm are given in Table 13.23.

Table 13.23. The best merit function value, the merit function mean value and the merit function standard deviation for all optimization algorithms. Only radii of curvature are variable.

	The best merit function value	The merit function mean value	The merit function standard deviation
Initial optical system	4.544	-	-
The DLS optimization	4.544	-	-
The ASSGA optimization	4.472	4.556	0.098
The ES EVOL optimization	4.462	4.462	0.000
The ES GRUP optimization	4.462	4.464	0.002
The ES REKO optimization	4.462	4.463	0.002
The ES KORR optimization	4.462	4.463	0.000

From Table 13.23 it can be seen that all optimizations have found optical systems with very similar merit functions. All evolution strategies optimization algorithms found the optical system with the best merit function. The merit function mean values are very close to the merit function best values which indicates that all optimizations have found almost identical optical systems. The explanation of this fact is that all optimization algorithms when variable were only radii of curvature had too few variable construction parameters to escape from local minimum and to search the optimization space successfully.

The second group of Cooke triplet optimizations is calculated when all radii of curvature and all separations between the optical surfaces are variable. The number of variable construction parameters is 11 (5 variable radii of curvature and 6 separations between the optical surfaces). Now the optimization algorithms have enough variable construction parameters to search the optimization space successfully and find optical systems with a significantly smaller merit function than the initial optical system merit function. The optical system with the best merit function found by any optimization algorithm, the merit function mean value and the merit function standard deviation calculated by any optimization algorithm are given in Table 13.24.

Table 13.24. The best merit function value, the merit function mean value and the merit function standard deviation for all optimization algorithms. The radii of curvature and the separations between the optical surfaces are variable.

	The best merit function value	The merit function mean value	The merit function standard deviation
Initial optical system	4.544	-	-
The DLS optimization	0.482	-	-
The ASSGA optimization	0.578	0.751	0.143
The ES EVOL optimization	0.468	0.476	0.008
The ES GRUP optimization	0.496	0.511	0.016
The ES REKO optimization	0.499	0.508	0.006
The ES KORR optimization	0.468	0.517	0.035

From Table 13.24 it can be seen that all optimizations have found the optical systems with similar merit functions. The ES EVOL and the ES KORR optimizations have found the optical system with the best merit function. All evolution strategies optimizations have the merit function mean values close to the best merit function value. The ASSGA optimizations have found the optical system with the worst merit function and have the greatest merit function mean value. This only means that 11 variable construction parameters are not sufficient to search the whole optimization space successfully.

The third group of the Cooke triplet optimizations is calculated when all radii of curvature, all separations between the optical surfaces and all glasses are variable. The number of variable construction parameters is 14 (5 variable radii of curvature, 6 separations between the optical surfaces and 3 glasses). The DLS optimization and all evolution strategies (the ES EVOL, the ES GRUP, the ES REKO and the ES KORR) expect that all variable construction parameters are continuous variables. The problem with the variable glasses is that they are not continuous variables but there are discrete variables that are chosen from the glass database. Each optimization algorithm overcomes this problem in a specific way. The ASSGA optimization does not have this problem. It can work with continuous and discrete variable construction parameters. The optical system with the best merit function found by any optimization algorithm, the merit function mean value and the merit function standard deviation calculated by any optimization algorithm are given in Table 13.25.

Table 13.25. The best merit function value, the merit function mean value and the merit function standard deviation for all optimization algorithms. The radii of curvature, the separations between the optical surfaces and the glasses are variable.

	The best merit function value	The merit function mean value	The merit function standard deviation
Initial optical system	4.544	-	-
The DLS optimization	0.963	-	-
The ASSGA optimization	0.398	0.773	0.337
The ES EVOL optimization	0.645	0.724	0.056
The ES GRUP optimization	0.759	0.889	0.112
The ES REKO optimization	0.589	0.642	0.058
The ES KORR optimization	0.413	0.786	0.245

From Table 13.25 it is clear that optimization algorithms which have a problem with the discrete variable optimization, have found optical systems with the worse merit function. This means that their optimization algorithms did not successfully search the optimization space and they generaly have found sub optimum minimums. The DLS optimization has the greatest problem with the variable glass because of the inadequate algorithm for glass variation, so the DLS optimization has found the optical system with the worst merit function.

The only optimization algorithm that can work without problems with the continuous and the discrete variable is the ASSGA optimization. It is natural that the ASSGA optimization has found the optical system with the best merit function. It must be noted that the evolutionary strategy ES KORR, although it is designed to work only with the continuous variables, has found the optical system with the merit function that is very close to the optical system merit function found by the ASSGA optimization. It is interesting to note that the optimization algorithms which have found the optical systems with the best merit functions have much larger values for the merit function mean value and the standard deviation. This means that they found an optical system with quite different merit functions. The optimization algorithm with the smallest merit function mean value and the standard deviation is the multimbered evolution strategy ES REKO.

Chapter 14

The Petzval objective optimizations

14.1 Introduction

The Petzval objective was the first photographic objective to be deliberately designed rather than being put together by an empirical selection of lenses out of box. As originally designed by Joseph Petzval it took the form of two doublets widely apart with a central aperture diaphragm. The basic form of the Petzval objective has been retained in many present-day applications, but detailed changes have been made in the doublets. In some instances, for example, the doublets have been replaced by doublets plus associated singlets. In other variants of the basic form a third doublet has been added.

The Petzval objective has an excellent correction for the spherical aberration and the coma, but because the Petzval sum is uncorrected the field angle is limited by the astigmatism to about $2\omega = 24°$ to $30°$. The main advantage of the Petzval objective is that it has a comparatively high aperture. This may be attributed to the fact that both doublets are positive. As a result the refractive power is spread among the doublets and each of them has a power less than that of the Petzval objective. The main defect of the Petzval objective is its uncorrected Petzval sum. The very features that permit its use at high apertures also lead to this high Petzval sum. The main application of the Petzval objective is the projection of 16 mm and 8 mm movie films, where its high aperture is eminently desirable and where its low field angle is tolerable.

14.2 The Petzval objective initial design and optimization goals

The initial design for the Petzval objective is a standard design with two doublets. It has the relative aperture $f/4$ and the field angle $2\omega = 20°$. There are two optimization goals for the following optimizations:

– the first optimization goal is obvious: optimize the Petzval objective in order to obtain the smallest possible minimum of the merit function. There is a large number of merit function local minima but the goal for the optimization algorithm is to try to find the smallest possible local minimum i.e. the global minimum. The merit function used in all optimizations is defined in Section 8.2

- the second optimization goal is to compare the optimization designs and to see how many different but good designs of the Petzval objective can be found by optimization algorithms.

All optimization example objectives: the Cooke triplet, the Petzval and the Double Gauss are optimized with the following variable parameters:

- the variable construction parameters;
- the definition of boundary conditions;
- the definition of aberration weights;
- the definition of the paraxial data that is to be controlled and the weighting factor for each controlled paraxial data;
- the variable optimization parameters.

All above stated variable parameters except the variable optimization parameters are connected with the selected optimization example and they are the same for all optimization algorithms. With the classical objectives like the Cooke triplet, the Petzval and the Double Gauss, the possible variable construction parameters are:

- all radii of curvature;
- all radii of curvature and all separations between the optical surfaces;
- all radii of curvature, all separations between the optical surfaces and all glasses.

The description of all other variable parameters is given in Section 13.2 and it will not be repeated here.

In this Chapter all optical systems, initial and optimized are represented only by the optical system merit function value. In this Chapter all optical systems, initial and optimized are represented only by the optical system merit function value. This is the most convenient representation of the optimization results. For more information about input data files and optimization results reader can contact author on email address darko@afrodita.rcub.bg.ac.yu.

14.3   DLS Optimizations

The mathematical theory of the DLS optimization is given in Chapter 2 and the implementation of this optimization algorithm in the program ADOS is given in Chapter 8. The DLS optimizations for the selected Petzval objective are calculated for the following conditions:

- all radii of curvature are variable;
- all radii of curvature and all separations between the optical surfaces are variable;
- all radii of curvature, all separations between the optical surfaces and all glasses are variable.

All DLS optimizations are calculated with the following damping algorithms:

- the additive damping;

– the multiplicative damping.

The first DLS optimizations of the Petzval objective are when all radii of curvature are variable. The Petzval objective consists of two separated doublets: one cemented with three possible variable radii of curvature and one air separated with four possible variable radii of curvature. Together there are seven possible variable radii of curvature. In the optimization only six radii of curvature are used and the seventh radius of curvature is used for keeping the effective focal length at a predefined value. The merit function of the initial Petzval objective and the optimized Petzval objectives are given in Table 14.1.

Table 14.1. The merit function values of the optimized Petzval objectives for different damping algorithms. Only radii of curvature are variable.

	Merit function
The initial optical system	70.149
The DLS optimization with the additive damping	12.383
The DLS optimization with the multiplicative damping	12.451

From Table 14.1 it can be seen that six variable radii of curvature are sufficient variable construction parameters for the optimization algorithm to search the optimization space successfully and find good local minimums. In fact the local minimum found by the DLS optimization with the additive damping represents the best-found local minimum by any optimization algorithm (the DLS, the ASSGA, the ES EVOL, the ES GRUP, the ES REKO. the ES KORR). Both damping algorithms have found optical systems with nearly identical merit function values.

The next series of the DLS optimizations are done when all radii of curvature and all separations between the optical surfaces are variable. Now the optimization algorithm has a possibility to change 12 construction parameters (6 radii of curvature and 6 separations between the optical surfaces). The merit functions of the initial optical system and the optimized optical systems are given in Table 14.2.

Table 14.2. The merit function values of the optimized Petzval objectives for different damping algorithms. The radii of curvature and the separations between optical surfaces are variable.

	Merit function
The initial optical system	70.149
The DLS optimization with the additive damping	10.604
The DLS optimization with the multiplicative damping	30.187

From Table 14.2 it can be seen that different damping algorithms have found quite different optical systems. The DLS optimization with the additive damping has found the optical system that has almost three times better merit function value than the optical system found by the same DLS optimization with the multiplicative damping. It is interesting to compare the optical system merit function values found by the DLS optimization when variable were only the radii of curvature and the radii of curvature and the separations between the optical surfaces. The number of variable construction parameters was in the former case 6 and in the latter case 12. The DLS optimization with additive damping has found, in both cases, optical systems with similar merit function values (12.383 and 10.604) while the number of variable construction parameters is doubled from 6 to 12. Even more interesting is the case of the DLS optimization when the multiplicative damping is used. When the radii of curvature were only variable construction parameters the optimization algorithm has found an optical system with the merit function value that was 2.4 times better than when the radii of curvature and the separations between optical surfaces were variable construction parameters. If all these facts are considered, it can be concluded that the optimum number of variable construction parameters for the DLS optimization is 6 i.e. when only radii of curvature are variable.

The last series of the DLS optimizations are done when all construction parameters are variable (all radii of curvature, all separations between the optical surfaces and all glasses). The total number of variable construction parameters is 16 (6 radii of curvature, 6 separations between the optical surfaces and 4 glasses). The DLS optimization algorithm has a problem with the discrete variable parameters like glasses because it expects the variable parameters to be continuous. The radii of curvature and the separations between the optical surfaces are continuous parameters but the glasses are discrete ones (i.e. there is a finite number of glasses in the manufactures glass database). A variable parameter ought to be continuous because the partial derivatives of aberrations with respect to the variable construction parameters are calculated by adding and subtracting increments from the variable parameter and calculating aberrations. The discrete parameter must be transformed to the continuous parameter and the whole optimization must be done with this continuous parameter. The DLS optimization algorithm usually finds a good local minimum with this continuous parameter, but when these continuous parameters are transformed to the real glasses from the glass database, the merit function of this real optical system may be and usually is quite different from the merit function of the optical system with the variable glasses as continuous parameters. The conversion from the glass continuous parameter to the nearest real glass is done so that:

– the difference between the calculated index of refraction and the real glass index of refraction is minimum;

– the difference between the calculated Abbe number and the real glass Abbe number is minimum.

In the author's opinion the DLS optimization with the variable glasses ought to be used only when it is absolutely necessary. The optical designer can select good glasses from the glass database and run the DLS optimization with only radii of curvature and separations between the optical surfaces variable.

The merit functions of the initial optical system and the optimized optical systems are given in Table 14.3. From Table 14.3 it is clear that the DLS optimization has a problem in searching the optimization space and finding good optical systems when the glasses are variable. When the damping algorithm was the additive damping the DLS optimization has found the optical system with the merit function value that is worse than when the glasses are not variable (11.120 and 10.604).

Table 14.3. The merit function values of the optimized Petzval objectives for different damping algorithms. The radii of curvature, the separations between optical surfaces and the glasses are variable.

	Merit function
The initial optical system	70.149
The DLS optimization with the additive damping	11.120
The DLS optimization with the multiplicative damping	-

When the damping algorithm was the multiplicative damping the DLS optimization did not manage to find an optical system that fulfils all boundary conditions. No adequate functioning of the multiplicative damping is evident in case when the radii of curvature and the separations between the optical surfaces were variable and especially in case when all construction parameters were variable (the radii of curvature, the separations between the optical surfaces and the glasses).

14.4 ASSGA optimizations

The mathematical theory of the ASSGA optimization is given in Chapter 3 and the implementation of this optimization algorithm in the program ADOS is given in Chapter 9. The ASSGA optimizations for the selected Petzval objective are calculated for the following conditions:

- all radii of curvature are variable;
- all radii of curvature and all separations between the optical surfaces are variable;
- all radii of curvature, all separations between the optical surfaces and all glasses are variable.

In all ASSGA optimizations the following strategy is accepted. Each series of the ASSGA optimizations is started with all default parameters. Two optimization parameter values: the population size and the starting values for genetic operator probabilities are independently varied in order to find the best combination.

Since the probabilistic events play an important role in the ASSGA optimization, it is necessary to run several optimizations to see a general behaviour of the optimization algorithm. The author decided to run five optimizations for each changed optimization parameter.

The first ASSGA optimizations of the Petzval objective are calculated when all radii of curvature are variable. There are six radii of curvature that can be varied. As in the DLS optimizations there are enough variable construction parameters and the ASSGA optimization algorithm can search the optimization space successfully and find optical systems with a significantly better merit function than the initial optical system merit function. The merit functions of the optimized optical systems for different population sizes and starting values for the genetic operator probabilities are given in Table 14.4.

Table 14.4. The merit functions of the optimized Petzval objectives for different population sizes and different starting values for the genetic operator probabilities. Only radii of curvature are variable.

	Merit function						
	1st opt.	2nd opt.	3rd opt.	4th opt.	5th opt.	mean value	stand. deviat.
Merit function of the initial Petzval objective is 70.149							
Population 300							
$AC = UC = FM = BM = SM = 20$	22.980	19.922	21.674	24.537	31.263	24.075	4.361
$AC = UC = 30, FM = 20, BM = SM = 10$	26.240	26.964	16.495	33.365	17.036	24.020	7.182
$AC = UC = 15, FM = 30, BM = SM = 20$	23.311	23.480	23.408	20.555	26.458	23.442	2.089
Population 400							
$AC = UC = FM = BM = SM = 20$	16.196	24.989	29.455	16.884	18.430	21.191	5.784
$AC = UC = 30, FM = 20, BM = SM = 10$	15.578	20.102	21.619	22.162	22.196	20.331	2.790
$AC = UC = 15, FM = 30, BM = SM = 20$	24.689	19.716	14.980	20.332	21.718	20.287	3.532
Population 500							
$AC = UC = FM = BM = SM = 20$	18.966	26.262	24.642	20.883	20.890	22.329	3.012
$AC = UC = 30, FM = 20, BM = SM = 10$	21.008	20.767	25.704	21.250	15.787	20.903	3.513
$AC = UC = 15, FM = 30, BM = SM = 20$	24.890	21.695	19.504	22.540	16.065	20.939	3.338

where:

AC is the average crossover;

UC is the uniform crossover;

FM is the floating-point number mutation;

BM is the big number creep;

SM is the small number creep.

From Table 14.4 it can be seen that the optimizations of the Petzval objective are calculated for the population sizes from 300 optical systems to 500 optical systems and for the following three sets of the starting values for the genetic operator probabilities:

- The first set is the default values for all genetic operator probabilities. The starting values for the genetic operator probabilities are the following: the average crossover, the uniform crossover, the floating point number mutation, the big number creep and the small number creep are equal to 20 ($AC = UC = FM = BM = SM = 20$).

- The second set is the set with larger values for the crossover probabilities than for the mutation probabilities. The starting values for the genetic operator probabilities are the following: the average crossover and the uniform crossover have the probability 30, the floating point number mutation has the probability 20, the big number creep and the small number creep have the probability 10 ($AC = UC = 30$, $FM = 20$, $BM = SM = 10$).

– The third set is the set with larger values for mutation probabilities than for the crossover probabilities. The starting values for the genetic operator probabilities are the following: the average crossover and the uniform crossover have the probability 15, the floating point number mutation has the probability 30, the big number creep and the small number creep have the probability 20 ($AC = UC = 30$, $FM = 20$, $BM = SM = 10$).

The best optical systems with the smallest merit functions are found when the population size was 400 optical systems. When the different sets of genetic operator probabilities are compared from Table 14.4 it can be seen that the set with larger crossover probabilities and the set with larger mutation probabilities have found optical systems with similar merit functions. The merit function mean value for these optical systems are better than for the optical systems found by the ASSGA optimization when all genetic operator probabilities are equal. The best optical system with the merit function 14.980 is found when the population was 400 optical systems and the starting values for the mutation probabilities were larger than for the crossover probabilities. It is interesting to notice that the DLS optimization has managed to find a better optical system with a smaller merit function than the ASSGA optimization.

The second set of the Petzval objective ASSGA optimizations is calculated when all radii of curvature and all separations between the optical surfaces are variable. The number of variable construction parameters is 12 (6 radii of curvature and 6 separations between the optical surfaces). The population size and the starting values for the genetic operator probabilities are independently varied. The genetic operator probabilities have three sets of values described on previous page. The merit functions of the optimized optical systems for different population sizes and starting values for the genetic operator probabilities are given in Table 14.5.

Table 14.5. The merit functions of the optimized Petzval objectives for different population sizes and the different starting values for the genetic operator probabilities. The radii of curvature and the separations between the optical surfaces are variable.

	Merit function						
	1st opt.	2nd opt.	3rd opt.	4th opt.	5th opt.	mean value	stand. deviat.
Merit function of the initial Petzval objective is 70.149							
$AC = UC = FM = BM = SM = 20$							
Population 200	26.660	14.127	34.142	19.484	12.836	21.450	8.943
Population 300	18.289	13.973	12.160	11.948	19.031	15.080	3.371
Population 400	13.836	18.610	22.747	12.572	12.327	16.018	4.535
Population 500	16.567	20.456	13.768	16.905	12.994	16.138	2.955
$AC = UC = 30, FM = 20, BM = SM = 10$							
Population 400	13.460	13.248	25.128	17.774	14.243	16.771	5.015
Population 500	12.871	11.806	17.315	13.850	12.750	13.718	2.137
Population 600	21.026	11.612	17.184	16.650	12.412	15.777	3.839

$AC = UC = 15$, $FM = 30$, $BM = SM = 20$							
Population 200	26.828	13.246	17.326	13.849	28.305	19.911	7.179
Population 300	19.672	16.307	11.892	14.299	14.119	15.258	2.921
Population 400	12.594	15.723	28.529	18.509	11.769	17.425	6.758

where:

AC is the average crossover;

UC is the uniform crossover;

FM is the floating-point number mutation;

BM is the big number creep;

SM is the small number creep.

From Table 14.5 it can be seen that the ASSGA optimization has found best optical systems when the population is 500 optical systems and the starting values for the genetic operator probabilities are greater for the crossover operators than for the mutation operators. This is a classical genetic algorithms scheme. For these condition the ASSGA has found the optical systems with the smallest merit function mean value and the standard deviation. Four out of five optimizations have found optical systems with the merit function from 11.806 to 13.850. The best optical system with the minimum merit function (11.612) is found when the population was 600 optical systems.

From the results presented in Table 14.5 very interesting phenomena can be seen. The basic assumption in the genetic algorithms is that a larger number of variable construction parameters needs a larger population size. This is logical because of the implicit parallelism of genetic algorithms, the ASSGA optimization with a larger population size can search better the optimization space. Here the situation is contrary to this basic assumption. When, only the radii of curvature are variable, the Petzval objective has six variable construction parameters and the optimum population size is 400 optical systems for any combination of starting values for the genetic operator probabilities. When, the radii of curvature and the separations between the optical surfaces are variable, the Petzval objective has twelve variable construction parameters (the number of variable parameters is doubled) and the optimum population size is 300 optical systems for two cases: when all starting values of genetic operator probabilities are equal and when the mutation operators have larger starting probabilities than the crossover operators. The criterion for the selection of the optimum population size is the minimum merit function mean value and the minimum merit function standard deviation. It is clear that because of described phenomena the ASSGA optimization was not able to find the best optical system with this combination of genetic operator probabilities. When the starting values for the crossover operator probabilities are greater than the mutation operators, the ASSGA optimization has found the optical systems with the merit function mean value 13.718 and in other two cases (all genetic operators have equal starting probability and the mutation operators have greater starting probability than the crossover operators) the ASSGA optimization has found the optical systems with the merit function mean value 15.080 and 15.258.

It is interesting to compare the best merit function value and the merit function mean value found by the ASSGA optimization when only the radii of curvature were variable and when the radii of curvature and the separations between the optical surfaces were variable. The number of variable construction parameters is doubled from six to twelve and the best merit function value is reduced for approximately 30% from 14.980 to 11.612. The merit function mean value is reduced for approximately 50% from 20.331 to 13.718.

The third set of the ASSGA optimizations of the Petzval objective is calculated when all radii of curvature, all separations between the optical surfaces and all glasses are variable. The number of variable construction parameters is 16 (6 radii of curvature, 6 separations between the optical surfaces and 4 glasses). The ASSGA optimization algorithm does not require all variables to be continuous, so the ASSGA optical system can work directly with glasses from the glass database. In fact the ASSGA optimization algorithm randomly selects a glass from the glass database. The ASSGA optimization algorithm has enough variable construction parameters to search the optimization space successfully and find the optical system with the best merit function. The ASSGA optimization has found much better optical systems with a smaller merit function when the glasses are variable than when the glasses are constant (not variable). Other optimization algorithms usually have found better optical systems when only the radii of curvature and the separations between the optical surfaces are variable. The following optimization parameters are independently varied: the population size and the starting values for the genetic operator probabilities. The genetic operator probabilities have three sets of values previously described. All other optimization parameters have default values. The merit functions of the optimized optical systems for different population sizes and starting values for the genetic operator probabilities are given in Table 14.6.

From Table 14.6 it can be seen that the ASSGA optimization has found optical systems with similar merit functions when the starting values for the genetic operator probabilities were equal and when the mutation operators had greater starting probabilities than the crossover operators. The ASSGA optimization managed to find many good optical systems with a merit function smaller than nine. The best optical systems found by the ASSGA optimization have a merit function smaller than eight. They are found when the population size was 600 optical systems and the starting values for genetic operator probabilities are equal and the population size was 900 optical systems and the mutation operators had greater starting probabilities than the crossover operators. In both cases the ASSGA optimization managed to find two optical systems with a merit function smaller than eight (7.805, 7.887 and 7.866, 7.835). The smallest merit function mean value and the merit function standard deviation have the ASSGA optimizations with the following parameters:

– the population is 600 optical systems and the starting values for all genetic operator probabilities are equal;
– the population is 800 optical systems and the mutation operators have greater starting probabilities than the crossover operators.

Table 14.6. The merit functions of the optimized Petzval objectives for different population sizes and different starting values for the genetic operator probabilities. The radii of curvature, the separations between the optical surfaces and the glasses are variable.

	Merit function						
	1st opt.	2nd opt.	3rd opt.	4th opt.	5th opt.	mean value	stand. deviat.
Merit function of the initial Petzval objective is 70.149							
$AC = UC = FM = BM = SM = 20$							
Population 500	8.667	8.665	11.700	9.463	9.032	9.505	1.270
Population 600	8.513	7.805	8.675	8.366	7.887	8.249	0.385
Population 700	9.035	8.155	9.048	8.551	8.446	8.647	0.388
$AC = UC = 30, FM = 20, BM = SM = 10$							
Population 500	10.034	9.381	11.245	10.415	8.416	9.898	1.068
Population 600	8.768	10.072	8.214	8.423	11.541	9.404	1.396
Population 700	9.551	9.118	8.352	11.617	9.174	9.562	1.228
$AC = UC = 15, FM = 30, BM = SM = 20$							
Population 600	9.336	8.357	8.734	8.655	8.851	8.787	0.357
Population 700	8.416	8.188	8.171	8.838	8.861	8.495	0.338
Population 800	8.773	8.620	8.096	8.006	8.291	8.357	0.331
Population 900	7.866	9.255	8.119	7.835	10.181	8.651	1.033

where:

AC is the average crossover;

UC is the uniform crossover;

FM is the floating-point number mutation;

BM is the big number creep;

SM is the small number creep.

When the starting probabilities for the crossover operators were greater than the mutation operators, the ASSGA optimization has found the optical systems with worse merit functions than in other optimization cases. It is interesting to note that when the radii of curvature and the separations between the optical surfaces were variable this combination of starting values for genetic operator probabilities have found the best optical systems with the smallest merit functions.

It is interesting to compare the number of variable parameters and the merit function values. When the radii of curvature and the separations between the optical surfaces are variable there are 12 variable construction parameters and the ASSGA optimization managed to find the optical systems with the merit function mean value 15.080. When the radii of curvature, the separations between the optical surfaces and the glasses are variable there are 16 variable construction parameters and the ASSGA optimization managed to find the optical systems with the merit function mean value 8.249, which is 80% increasement.

14.5 Evolution strategies optimizations

The evolution strategies are global optimization algorithms based on the theory of evolution and natural selection developed for the optimization of complex technical systems. In this monograph the mathematical theory for four different evolution strategies is given in Chapter 4 and their implementation in the program ADOS is given in Chapters 10 – 12. Here will be presented the optimizations of the selected Petzval objective with the following evolution strategies:
— the two membered evolution strategy ES EVOL;
— the multimembered evolution strategy ES GRUP;
— the multimembered evolution strategy ES REKO;
— the multimembered evolution strategy ES KORR.

14.5.1   Two membered evolution strategy ES EVOL optimizations

The two-membered evolution strategy ES EVOL represents the most simple evolution strategy. The ES EVOL optimizations for the selected Petzval objective are calculated for the following conditions:
— all radii of curvature are variable;
— all radii of curvature and all separations between the optical surfaces are variable;
— all radii of curvature, all separations between the optical surfaces and all glasses are variable.

The two membered evolution strategy ES EVOL has a small number of optimization parameters and all parameters are kept at default values.

The first ES EVOL optimizations of the Petzval objective are calculated when all radii of curvature are variable. There are six radii of curvature that can be varied. As in the DLS and the ASSGA optimizations there are enough variable construction parameters and the ES EVOL optimization algorithm can search optimization space successfully and find the optical systems with a significantly better merit function than the initial optical system merit function. The merit functions of the optimized optical systems are given in Table 14.7.

Table 14.7. The merit functions of the optimized Petzval objectives. Only radii of curvature are variable.

	Merit function
The 1st optimization	13.335
The 2nd optimization	13.356
The 3rd optimization	13.388
The 4th optimization	13.420
The 5th optimization	13.414
The merit function mean value	13.383
The merit function standard deviation	0.037

The ES EVOL optimizations have found the optical systems with very similar merit functions from 13.335 to 13.420. The merit function mean value is 13.383 and the merit function standard deviation is 0.037. These results show that all ES EVOL optimizations have found almost the same local minimum. It is interesting to compare optical system merit functions found by various optimization algorithms. The ES EVOL optimization has found the optical system with a better merit function than the ASSGA optimization and with a worse merit function than the DLS optimization.

The second group of the ES EVOL optimizations of the Petzval objective is calculated when all radii of curvature and all separations between the optical surfaces are variable. The number of variable construction parameters is 12 (6 radii of curvature and 6 separations between the optical surfaces). The merit functions of the optimized optical systems are given in Table 14.8.

Table 14.8. The merit functions of the optimized Petzval objectives. The radii of curvature and the separations between the optical surfaces are variable.

	Merit function
The 1st optimization	11.127
The 2nd optimization	11.195
The 3rd optimization	11.222
The 4th optimization	11.237
The 5th optimization	11.101
The merit function mean value	11.176
The merit function standard deviation	0.060

The ES EVOL optimizations have found optical systems with very similar merit functions from 11.101 to 11.237. The merit function mean value is 11.176 and the merit function standard deviation is 0.060. These results show that all ES EVOL optimizations have found almost the same local minimum. It is interesting to note that the number of variable construction parameters is doubled from six to twelve and the optical system merit function is reduced for only approximately 20% from 13.335 to 11.101. If the optical system merit functions found by various optimization algorithms are compared then the ES EVOL optimization has found the optical system with a better merit function than the ASSGA optimization and with a worse merit function than the DLS optimization.

The third group of the ES EVOL optimizations of the Petzval objective is calculated when all radii of curvature, all separations between the optical surfaces and all glasses are variable. The number of variable construction parameters is 16 (6 radii of curvature, 6 separations between the optical surfaces and 4 glasses). The evolution strategies are originally designed for continuous variable optimization. The problem rises because the variable glasses are not continuous parameters. They are discrete parameters selected from the database with the finite number of different glasses. In the optimization the glass variation (mutation) consists of two steps. In the first step glasses are treated as all other variable construction parameters. The glass index of refraction and the Abbe number are mutated as other continuous parameters. In the second step the glass database is searched to find the corresponding glass to new obtained values for the glass index of refraction and the Abbe number. The new glass must be of the same type meaning if the starting glass is a crown glass then the new changed (mutated) glass must be also a crown glass. If the mutated glass is not the same type as the starting glass then the mutation is rejected and another one is calculated. This method of variable glass mutation gives a large number of rejected mutations and the 1/5 rule is not fulfilled any more. The 1/5 rule states that, in average, one out of five mutations ought to be successful meaning it ought to produce an optical system that has a better merit function than the starting optical system. The consequence of not fulfilling the 1/5 rule is very slow convergence to the minimum which is not the best possible local minimum and a large number of executed mutations with a small number of successful mutations. The additional problem in the Petzval objective optimization is that the selected Petzval objective does not fulfil all boundary conditions. This means that the ES EVOL optimization must be executed two times. The first time is to find the starting Petzval objective that fulfils all boundary conditions and the second time is to optimize this new obtained Petzval objective. In the first ES EVOL optimization the variable construction parameters are freely changing in order to find an optical system that fulfils all boundary conditions. Usually, the optical system that fulfils all boundary conditions has the merit function much larger than the starting optical system. To make situation even worse, the variable optical glasses are usually exotic ones. These glasses are even more difficult to change than standard glasses. The merit functions of the optimized optical systems are given in Table 14.9.

Table 14.9. The merit functions of the optimized Petzval objectives. The radii of curvature, the separations between the optical surfaces and the glasses are variable.

	Merit function
The 1st optimization	12.849
The 2nd optimization	11.573
The 3rd optimization	13.258
The 4th optimization	12.260
The 5th optimization	11.324
The merit function mean value	12.253
The merit function standard deviation	0.820

From Table 14.9 it is obvious that the ES EVOL optimization has great problems to find good optical systems with small merit functions. The ES EVOL optimization has found the optical system with the merit functions from 11.324 to 13.258. The merit function mean value is 12.253 and the merit function standard deviation is 0.820. This large standard deviation is the first sign that the optimization algorithm has difficulties in finding good local minimums. When the radii of curvature and the separations between the optical surfaces were only variable the standard deviations were much smaller (0.037 and 0.060). Furthermore, optical systems have larger merit functions than the merit functions of optical systems found by the same optimization algorithm when the radii of curvature and the separations between the optical surfaces were only variable.

14.5.2   Multimembered evolution strategy ES GRUP optimizations

The multimembered evolution strategy ES GRUP represents the further development of ideas from the two membered evolution strategy ES EVOL. The ES GRUP optimizations for the selected Petzval objective are calculated for the following conditions:
- all radii of curvature are variable;
- all radii of curvature and all separations between the optical surfaces are variable;
- all radii of curvature, all separations between the optical surfaces and all glasses are variable.

The multimembered evolution strategy ES GRUP has a small number of optimization parameters and all parameters are kept at default values.

The first ES GRUP optimizations of the Petzval objective are calculated when all radii of curvature are variable. There are six radii of curvature that can be varied. As in all preceding optimizations there are enough variable construction parameters and the ES GRUP optimization algorithm can search optimization space successfully and find optical systems with a significantly better merit function than the initial optical system merit function. The merit functions of the optimized optical systems are given in Table 14.10.

Table 14.10. The merit functions of the optimized Petzval objectives. Only radii of curvature are variable.

	Merit function
The 1st optimization	13.823
The 2nd optimization	13.423
The 3rd optimization	13.447
The 4th optimization	13.567
The 5th optimization	13.498
The merit function mean value	13.552
The merit function standard deviation	0.161

The ES GRUP optimizations have found the optical systems with similar merit functions from 13.423 to 13.823. The merit function mean value is 13.552 and the merit function standard deviation is 0.161. The ES GRUP optimization that represents the further improvement of the ES EVOL optimization has found optical systems with worse merit functions than the ES EVOL optimization. The difference is small but significant. The best merit function values are 13.335 for the ES EVOL optimization and 13.423 for the ES GRUP optimization. The merit function mean values are 13.383 for the ES EVOL optimization and 13.552 for the ES GRUP optimization. It is interesting to compare the ES GRUP optimization with the other optimization algorithms (the DLS optimization and the ASSGA optimization). The ES GRUP optimization is similar to the ES EVOL optimization and it has found the optical system with a better merit function than the ASSGA optimization and with a worse merit function than the DLS optimization.

The second group of the ES GRUP optimizations of the Petzval objective is calculated when all radii of curvature and all separations between the optical surfaces are variable. The number of variable construction parameters is 12 (6 radii of curvature and 6 separations between the optical surfaces). The merit functions of the optimized optical systems are given in Table 14.11.

Table 14.11. The merit functions of the optimized Petzval objectives. The radii of curvature and the separations between the optical surfaces are variable.

	Merit function
The 1st optimization	11.272
The 2nd optimization	11.471
The 3rd optimization	11.008
The 4th optimization	11.584
The 5th optimization	11.397
The merit function mean value	11.346
The merit function standard deviation	0.221

The ES GRUP optimizations have found the optical systems with similar merit functions from 11.008 to 11.584. The merit function mean value is 11.346 and the merit function standard deviation is 0.221. It is interesting to note that the number of variable construction parameters is doubled from six to twelve and the optical system merit function is reduced for only approximately 20% from 13.423 to 11.008. The ES GRUP optimization that represents the further improvement of the ES EVOL optimization has found, as expected, a better optical system with the best merit function than the ES EVOL optimization, but it has a worse merit function mean value than the ES EVOL optimization. The difference is small but significant. The best merit function values are 11.008 for the ES GRUP optimization and 11.101 for the ES EVOL optimization. The merit function mean values are 11.346 for the ES GRUP optimization and 11.176 for the ES EVOL optimization. These results show that the ES GRUP optimization has searched the optimization space better and has found different local minimums. It is interesting to compare the ES GRUP optimization with the other optimization algorithms (the DLS optimization and the ASSGA optimization). The ES GRUP optimization is similar to the ES EVOL optimization and it has found the optical systems with a better merit function than the ASSGA optimization and with a worse merit function than the DLS optimization.

The third group of the ES GRUP optimizations of the Petzval objectives is calculated when all radii of curvature, all separations between the optical surfaces and all glasses are variable. The number of variable construction parameters is 16 (6 radii of curvature, 6 separations between the optical surfaces and 4 glasses). The problem with variable glasses is the same as in the ES EVOL optimization described in Section 14.5.1. The consequence of this problem is a very slow convergence to the minimum which is not the best possible local minimum and a large number of executed mutations with a small number of successful mutations. The merit functions of the optimized optical systems are given in Table 14.12.

Table 14.12. The merit functions of the optimized Petzval objectives. The radii of curvature, the separations between the optical surfaces and the glasses are variable.

	Merit function
The 1st optimization	10.249
The 2nd optimization	10.967
The 3rd optimization	13.828
The 4th optimization	9.510
The 5th optimization	12.522
The merit function mean value	11.415
The merit function standard deviation	1.749

From Table 14.12 it can be seen that the ES GRUP optimization has found different optical systems with the merit functions from 9.510 to 13.828. The merit function mean value is 11.415 and the merit function standard deviation is 1.749. The comparison of the optical systems found by the ES GRUP optimization when the radii of curvature and the separations between the optical surfaces were only variables and when all construction parameters were variable shows that the ES GRUP optimization with all variable construction parameters has found a better optical system with the best merit function.

The ES GRUP optimization that represents the further improvement of the ES EVOL optimization has found, as expected, optical systems with better merit functions than the ES EVOL optimization. The best merit function values are 9.510 for the ES GRUP optimization and 11.324 for the ES EVOL optimization. The merit function mean values are 11.415 for the ES GRUP optimization and 12.253 for the ES EVOL optimization. It is interesting to compare the ES GRUP optimization with the other optimization algorithms (the DLS optimization and the ASSGA optimization). The ES GRUP optimization has found the optical systems with a better merit function than the DLS optimization and with a worse merit function than the ASSGA optimization.

14.5.3   Multimembered evolution strategy ES REKO optimizations

The multimembered evolution strategy ES REKO represents the further development of ideas from the two membered evolution strategy ES EVOL and the multimebered evolution strategy ES GRUP. The ES REKO optimizations for the selected Petzval objective are calculated for the following conditions:

- all radii of curvature are variable;
- all radii of curvature and all separations between the optical surfaces are variable;

– all radii of curvature, all separations between the optical surfaces and all glasses are variable.

The multimembered evolution strategy ES REKO has the same set of optimization parameters as the ES REKO optimization and all parameters are kept at default values.

The first ES REKO optimizations of the Petzval objective are calculated when all radii of curvature are variable. There are six radii of curvature that can be varied. As in all preceding optimizations there are enough variable construction parameters and the ES GRUP optimization algorithm can search optimization space successfully and find optical systems with a significantly better merit function than the initial optical system merit function. The merit functions of the optimized optical systems are given in Table 14.13.

Table 14.13. The merit functions of the optimized Petzval objectives. Only radii of curvature are variable.

	Merit function
The 1st optimization	13.168
The 2nd optimization	13.161
The 3rd optimization	13.199
The 4th optimization	13.182
The 5th optimization	13.186
The merit function mean value	13.179
The merit function standard deviation	0.015

The ES REKO optimizations have found almost identical optical systems with very similar merit functions from 13.161 to 13.199. The merit function mean value is 13.179 and the merit function standard deviation is 0.015. The ES REKO optimization has found the optical systems with better merit functions than either the ES EVOL or the ES GRUP optimization. The difference is small but significant. The best merit function values are 13.161 for the ES REKO optimization, 13.335 for the ES EVOL optimization and 13.423 for the ES GRUP optimization. The merit function mean values are 13.179 for the ES REKO optimization, 13.383 for the ES EVOL optimization and 13.552 for the ES GRUP optimization. It is interesting to compare the ES REKO optimization with the other optimization algorithms (the DLS optimization and the ASSGA optimization). The ES REKO optimization is similar to the other evolution strategies and it has found an optical system with a better merit function than the ASSGA optimization and with a worse merit function than the DLS optimization.

The second group of the ES REKO optimizations of the Petzval objective is calculated when all radii of curvature and all separations between the optical surfaces are variable. The number of variable construction parameters is 12 (6 radii of curvature and 6 separations between the optical surfaces). The merit functions of the optimized optical systems are given in Table 14.14.

Table 14.14. The merit functions of the optimized Petzval objectives. The radii of curvature and the separations between the optical surfaces are variable.

	Merit function
The 1st optimization	10.842
The 2nd optimization	10.726
The 3rd optimization	11.200
The 4th optimization	10.738
The 5th optimization	11.260
The merit function mean value	10.953
The merit function standard deviation	0.258

The ES REKO optimizations have found the optical systems with similar merit functions from 10.726 to 11.260. The merit function mean value is 10.953 and the merit function standard deviation is 0.258. It is interesting to note that the number of variable construction parameters is doubled from six to twelve and the optical system merit function is reduced for only approximately 23% from 13.161 to 10.726. The ES REKO optimization that represents the further improvement of the ES EVOL and the ES GRUP optimization has found, as expected, optical systems with better merit functions than the ES EVOL and the ES GRUP optimization. The best merit function values are 10.726 for the ES REKO optimization, 11.008 for the ES GRUP optimization and 11.101 for the ES EVOL optimization. The merit function mean values are 10.953 for the ES REKO optimization, 11.346 for the ES GRUP optimization and 11.176 for the ES EVOL optimization. It is interesting to compare the ES REKO optimization with the other optimization algorithms (the DLS optimization and the ASSGA optimization). The ES REKO optimization has found optical systems with merit functions very close to the optical system merit function found by the DLS optimization. As other evolution strategies optimizations the ES REKO has found optical systems with better merit functions than the ASSGA optimization.

The third group of the ES REKO optimizations of the Petzval objectives is calculated when all radii of curvature, all separations between the optical surfaces and all glasses are variable. The number of variable construction parameters is 16 (6 radii of curvature, 6 separations between the optical surfaces and 4 glasses). The problem with variable glasses is the same as in all evolution strategies optimizations and it is described in Section 14.5.1. The consequence of this problem is a very slow convergence to the minimum which is not the best possible local minimum and a large number of executed mutations with a small number of successful mutations. The merit functions of the optimized optical systems are given in Table 14.15.

From Table 14.15 it can be seen that the ES REKO optimization has found different optical systems with the merit functions from 8.918 to 10.882. The merit function mean value is 10.071 and the merit function standard deviation is 0.867. The comparison of the optical systems found by the ES REKO optimization when the radii of curvature and the separations between the optical surfaces were only variables and when all construction parameters were variable shows that the ES

REKO optimization with all variable construction parameters has found the optical systems with approximately 20% better merit functions.

Table 14.15. The merit functions of the optimized Petzval objectives. The radii of curvature, the separations between the optical surfaces and the glasses are variable.

	Merit function
The 1st optimization	10.882
The 2nd optimization	10.684
The 3rd optimization	8.918
The 4th optimization	10.485
The 5th optimization	9.394
The merit function mean value	10.071
The merit function standard deviation	0.867

The ES REKO optimization that represents the further improvement of the ES EVOL and the ES GRUP optimization has found, as expected, optical systems with better merit functions than other evolution strategies. The best merit function values are 8.918 for the ES REKO optimization, 9.510 for the ES GRUP optimization and 11.324 for the ES EVOL optimization. The merit function mean values are 10.071 for the ES REKO optimization, 11.415 for the ES GRUP optimization and 12.253 for the ES EVOL optimization. It is interesting to compare the ES REKO optimization with the other optimization algorithms (the DLS optimization and the ASSGA optimization). The ES REKO optimization is similar to other evolution strategies and it has found an optical system with a better merit function than the DLS optimization and with a worse merit function than the ASSGA optimization.

14.5.4   Multimembered evolution strategy ES KORR optimizations

The multimembered evolution strategy ES KORR is the most developed and the most complex evolution strategy. The ES KORR optimizations for the selected Petzval objective are calculated for the following conditions:

– all radii of curvature are variable;
– all radii of curvature and all separations between the optical surfaces are variable;
– all radii of curvature, all separations between the optical surfaces and all glasses are variable.

All optimization parameters in the multimembered evolution strategy ES KORR are kept at default values.

The first ES KORR optimizations of the Petzval objective are calculated when all radii of curvature are variable. There are six radii of curvature that can be varied. As in all preceding optimizations there are enough variable construction parameters and the ES KORR optimization algorithm can search optimization space successfully and find optical systems with a significantly better merit function than the initial optical system merit function. The merit functions of the optimized optical systems are given in Table 14.16.

Table 14.16. The merit functions of the optimized Petzval objectives. Only radii of curvature are variable.

	Merit function
The 1st optimization	13.267
The 2nd optimization	13.612
The 3rd optimization	14.773
The 4th optimization	13.297
The 5th optimization	14.281
The merit function mean value	13.846
The merit function standard deviation	0.659

The ES KORR optimization has found the optical systems with the different merit functions from 13.267 to 14.773. Two out of five optical systems have small merit functions close to the best merit functions found by the evolution strategies and two out of five optical system have the largest merit functions found by the evolution strategies. If the merit function mean values are compared then the ES KORR optimization has the worst value in all evolution strategies. If only the best merit functions are considered then the ES KORR optimization has found the optical system with the merit function better than the ES EVOL and the ES GRUP optimization and worse than the ES REKO optimization.

The second group of the ES KORR optimizations of the Petzval objective is calculated when all radii of curvature and all separations between the optical surfaces are variable. The number of variable construction parameters is 12 (6 radii of curvature and 6 separations between the optical surfaces). The merit functions of the optimized optical systems are given in Table 14.17.

Table 14.17. The merit functions of the optimized Petzval objectives. The radii of curvature and the separations between the optical surfaces are variable.

	Merit function
The 1st optimization	10.349
The 2nd optimization	10.838
The 3rd optimization	11.281
The 4th optimization	10.782
The 5th optimization	12.663
The merit function mean value	11.183
The merit function standard deviation	0.891

The ES KORR optimization has found the optical systems with the different merit functions from 10.349 to 12.663. Three out of five optical systems are exceptionally good optical systems with one optical system which has the best merit function (10.349) found by any optimization algorithm. If the increment of variable construction parameters and the reduction of optical system merit function is considered then in the ES KORR optimization for doubling the variable construction parameters from six to twelve the optical system merit function is reduced for approximately 28% from 13.267 to 10.349. The ES KORR optimization that represents the most complex and the most developed evolution strategy has found, as expected, the optical system with the best merit function but has the merit function mean value worse then the ES REKO optimization. The merit function mean values are 11.183 for the ES KORR optimization, 10.953 for the ES REKO optimization, 11.346 for the ES GRUP optimization and 11.188 for the ES EVOL optimization.

The third group of the ES KORR optimizations of the Petzval objectives is calculated when all radii of curvature, all separations between the optical surfaces and all glasses are variable. The number of variable construction parameters is 16 (6 radii of curvature, 6 separations between the optical surfaces and 4 glasses). The problem with variable glasses is the same as in all evolution strategies optimizations and it is described in Section 14.5.1. The consequence of this problem is a very slow convergence to the minimum, which is not the best possible local minimum and a large number of executed mutations with a small number of successful mutations. The ES KORR as the most elaborated evolution strategy has the possibility to rotate mutation hyperellipsoid. The basic idea about rotation of mutation hyperellipsoid and its influence on variable glasses is given in Section 13.5.4. In the case of the Petzval objective, the ES KORR optimizations are calculated for both cases: when the mutation hyperellipsoid can rotate and cannot rotate. The merit functions of the optimized optical systems are given in Table 14.18.

Table 14.18. The merit functions of the optimized Petzval objectives. The radii of curvature, the separations between the optical surfaces and the glasses are variable.

	The mutation hyperellipsoid can rotate	The mutation hyperellipsoid cannot rotate
The 1st optimization	10.282	11.145
The 2nd optimization	9.680	10.884
The 3rd optimization	10.417	11.119
The 4th optimization	11.130	10.302
The 5th optimization	10.753	9.914
The merit function mean value	10.452	10.673
The merit function standard deviation	0.542	0.543

From Table 14.18 it can be seen that the ES KORR optimization in both cases has found the optical systems with the similar merit functions. The ES KORR optimization with the mutation hyperellipsoid rotation has found the optical system with the best merit function 9.680 and the merit function mean value 10.452. The ES KORR optimization without mutation hyperellipsoid rotation has found the optical system with the best merit function 9.914 and the merit function mean value 10.673. The comparison of the optical systems found by the ES KORR optimization when the radii of curvature and the separations between the optical surfaces were only variables and when all construction parameters were variable shows that the ES KORR optimization with all variable construction parameters has found the optical systems with only approximately 7% better merit functions. This small reduction in the merit function value can be explained by bad influence of glass variation because the glasses are discrete variables and evolution strategies are designed for continuous variables. The ES KORR optimization is similar to other evolution strategies and it has found the optical system with a better merit function than the DLS optimization and with a worse merit function than the ASSGA optimization.

14.6   Comparison of the Petzval objective designs by various optimization methods

All optimization methods: the DLS optimization, the ASSGA optimization, the two membered evolution strategy ES EVOL optimization, the multimembered evolution strategies ES GRUP, ES REKO and ES KORR optimizations have calculated various Petzval objectives with the following variable construction parameters:

− all radii of curvature;
− all radii of curvature and all separations between the optical surfaces;
− all radii of curvature and all separations between the optical surfaces and all glasses.

The first group of the Petzval objective optimizations is calculated when all radii of curvature are variable. There are six radii of curvature that can be varied. The optical system with the best merit function found by any optimization algorithm, the merit function mean value and the merit function standard deviation calculated for any stochastic optimization algorithm are given in Table 14.19.

Table 14.19. The best merit function value, the merit function mean value and the merit function standard deviation for all optimization algorithms. Only radii of curvature are variable.

	The best merit function value	The merit function mean value	The merit function standard deviation
Initial optical system	70.149	-	-
The DLS optimization	12.383	-	-
The ASSGA optimization	14.980	20.287	3.532
The ES EVOL optimization	13.335	13.383	0.037
The ES GRUP optimization	13.423	13.552	0.161
The ES REKO optimization	13.161	13.179	0.015
The ES KORR optimization	13.267	13.846	0.659

From Table 14.19 it can be seen that all optical systems optimized by various optimization methods represent significant improvement to the starting optical system. All optimized optical systems have the merit functions more than five times smaller than the starting optical system. The optical system with the smallest merit function is found by the DLS optimization. It has an approximately 10% smaller merit function than the merit function of optical systems found by other optimization methods. All evolution strategies have found optical systems with very similar merit functions. Among the evolution strategies the best performing strategy is the ES REKO which:

- has found the optical system with the minimum merit function;
- has the minimum merit function mean value;
- has the minimum merit function standard deviation value.

The ASSGA optimization has found the optical systems with the biggest merit functions. This clearly shows that only six variable radii of curvature are not enough for the ASSGA optimization to search the optimization space successfully.

The second group of the Petzval objective optimizations is calculated when all radii of curvature and all separations between the optical surfaces are variable. The number of variable construction parameters is 12 (6 radii of curvature and 6 separations between the optical surfaces). The optical system with the best merit function found by any optimization algorithm, the merit function mean value and the merit function standard deviation calculated for any stochastic optimization algorithm are given in Table 14.20.

Table 14.20. The best merit function value, the merit function mean value and the merit function standard deviation for all optimization algorithms. The radii of curvature and the separations between the optical surfaces are variable.

	The best merit function value	The merit function mean value	The merit function standard deviation
Initial optical system	70.149	-	-
The DLS optimization	10.604	-	-
The ASSGA optimization	11.612	15.777	3.839
The ES EVOL optimization	11.101	11.176	0.060
The ES GRUP optimization	11.008	11.346	0.221
The ES REKO optimization	10.726	10.953	0.258
The ES KORR optimization	10.349	11.183	0.891

From Table 14.20 it can be seen that all optical systems optimized by various optimization methods represent significant improvement to the starting optical system. All optimized optical systems have the merit functions more than six times smaller than the starting optical system. The optical system with the smallest merit function is found by the ES KORR optimization. The DLS optimization and the ES REKO optimization have found the optical systems with very similar merit functions. Other evolution strategies and the ASSGA optimization have found similar optical systems that have a slightly worse merit function. The ASSGA optimization has found the optical systems with the worst merit functions but the difference between the optical system merit functions found by the ASSGA optimization and other optimizations is reduced very much. The number of variable parameters, which are now doubled from six to twelve, is barely enough to search the optimization space successfully.

The third group of the Petzval objective optimizations is calculated when all radii of curvature, all separations between the optical surfaces and all glasses are variable. The number of variable construction parameters is 16 (6 radii of curvature, 6 separations between the optical surfaces and 4 glasses). The DLS optimization and all evolution strategies (the ES EVOL, the ES GRUP, the ES REKO and the ES KORR) expect that all variable construction parameters are continuous variables. The problem with the variable glasses is that they are not continuous variables but there are discrete variables that are chosen from the glass database. Each optimization algorithm overcomes this problem in a specific way. The ASSGA optimization does not have this problem. It can work with continuous and discrete variable construction parameters. The optical system with the best merit function found by any optimization algorithm, the merit function mean value and the merit function standard deviation calculated by any optimization algorithm are given in Table 14.21.

Table 14.21. The best merit function value, the merit function mean value and the merit function standard deviation for all optimization algorithms. The radii of curvature, the separations between the optical surfaces and the glasses are variable.

	The best merit function value	The merit function mean value	The merit function standard deviation
Initial optical system	70.149	-	-
The DLS optimization	11.120	-	-
The ASSGA optimization	7.805	8.249	0.385
The ES EVOL optimization	11.324	12.253	0.820
The ES GRUP optimization	9.510	11.415	1.749
The ES REKO optimization	8.918	10.071	0.867
The ES KORR optimization	9.680	10.452	0.542

From Table 14.21 it can be seen that all optical systems optimized by various optimization methods represent significant improvement to the starting optical system. All optimized optical systems have the merit functions from six times to nine times smaller than the starting optical system. The optical system with the smallest merit function is found by the ASSGA optimization because the ASSGA optimization has finally enough variable parameters and because it has no problem with the variable glasses, which are discrete parameters. Other optimization methods: the DLS optimization and all evolution strategies expect that all variable parameters are continuous. Each of these optimization methods has a special system for dealing with discrete parameters. Among the optimization methods, which expect the continuous variable parameters, the best optical systems are found by the ES REKO optimization. The ES REKO optimization has found the optical system with the smallest merit function and it also has the smallest merit function mean value. The optical systems with the worst merit function are found by the DLS optimization and the ES EVOL optimization because they have the biggest problems with variable glasses as discrete parameters.

Chapter 15

The Double Gauss objective optimizations

15.1 Introduction

The mathematician Gauss once suggested that a telescope objective could be made with two meniscus – shaped elements. The idea was that such a system would be free from spherochromatism. However, this arrangement has other serious disadvantages and it has not been used in any large telescope. After lot of experiments it was recognised that two such objectives mounted symmetrically about the central stop might make a good photographic objective. One of the most popular Double Gauss objectives is designed in the famous German optical company Zeiss and it was called Biotar. The Biotar objective in its basic form consists of a single positive lens, followed by a negative meniscus doublet that is concave to the rear, followed by another negative meniscus doublet that is concave to the front, and a final positive single lens. The thicknesses of meniscus components are approximately equal and the arrangement of refractive indices is often symmetric with $n_1 \approx n_6$, $n_2 \approx n_5$ and $n_3 \approx n_4$. This is an exceedingly powerful design form, and many high performance objectives are modifications or elaborations of this type. If the back focal length is made short and the elements are strongly curved about the central stop, fairly wide fields may be covered. Conversely, a long system with flatter curves will cover a narrow field at high aperture.

The Double Gauss objective is used in television and photographic cameras and it has excellent correction of all aberrations. The only problem that the Double Gauss objective has is the correction of the rim ray aberrations. Various efforts and modifications of the basic design have been proposed in order to correct these rim ray aberrations:

- the first and the most interesting modification of the basic design is the increment of the central separation between the two groups of lenses;
- the other modification of the basic design is adding one more lens at the end of the objective.

15.2 The Double Gauss objective initial design and optimization goals

The initial design for the Double Gauss objective is the standard Zeiss Biotar in its basic form. The Biotar consists of six lenses and it has relative aperture $f/2$ and the field angle $2\omega = 20°$. There are two optimization goals for the following optimizations:

– the first optimization goal is obvious: optimize the Double Gauss objective in order to obtain the smallest possible minimum of the merit function. There is a large number of merit function local minima but the goal for the optimization algorithm is to try to find the smallest possible local minimum i.e. the global minimum. The merit function used in all optimizations is defined in Section 8.2

– the second optimization goal is to compare the optimization designs and to see how many different but good designs of the Double Gauss objective can be found by optimization algorithms.

All optimization example objectives: the Cooke triplet, the Petzval and the Double Gauss are optimized with the following variable parameters:

– the variable construction parameters;

– the definition of boundary conditions;

– the definition of aberration weights;

– the definition of the paraxial data that is to be controlled and the weighting factor for each controlled paraxial data;

– the variable optimization parameters.

All above stated variable parameters except the variable optimization parameters are connected with the selected optimization example and they are the same for all optimization algorithms. With the classical objectives like the Cooke triplet, the Petzval and the Double Gauss possible variable construction parameters are:

– all radii of curvature;

– all radii of curvature and all separations between the optical surfaces;

– all radii of curvature, all separations between the optical surfaces and all glasses.

The description of all other variable parameters is given in Section 13.2 and it will not be repeated here.

In this Chapter all optical systems, initial and optimized are represented only by the optical system merit function value. This is the most convenient representation of the optimization results. For more information about input data files and optimization results reader can contact author on email address darko@afrodita.rcub.bg.ac.yu.

15.3   DLS Optimizations

The mathematical theory of the DLS optimization is given in Chapter 2 and the implementation of this optimization algorithm in the ADOS program is given in Chapter 8. The DLS optimizations for the selected Double Gauss objective are calculated for the following conditions:

– all radii of curvature are variable;

– all radii of curvature and all separations between the optical surfaces are variable;

– all radii of curvature, all separations between the optical surfaces and all glasses are variable.

All DLS optimizations are calculated with the following damping algorithms:

– the additive damping;
– the multiplicative damping.

The first DLS optimizations of the Double Gauss objective are when all radii of curvature are variable. The Double Gauss objective consists of two single lenses and two cemented doublets. Each single lens has two possible variable radii of curvature and each cemented doublet has three possible variable radii of curvature. Together there are ten possible variable radii of curvature. In the optimization only nine radii of curvature are used and the tenth radius of curvature is used for keeping the effective focal length at a predefined value. The merit function of the initial Double Gauss objective and the optimized Double Gauss objectives are given in Table 15.1.

Table 15.1. The merit function values of the optimized Double Gauss objectives for different damping algorithms. Only radii of curvature are variable.

	Merit function
The initial optical system	21.387
The DLS optimization with the additive damping	3.044
The DLS optimization with the multiplicative damping	4.030

From Table 15.1 it can be seen that nine variable radii of curvature are sufficient variable construction parameters for the optimization algorithm to search the optimization space successfully and find good local minimums. The merit function of the optimized optical systems is seven times better than the merit function for the initial optical system in the case of additive damping and approximately five times better in the case of multiplicative damping. The DLS optimization with the additive damping has found a better local minimum which is close to the best local minimum found by the evolution strategy ES REKO optimization.

The next series of DLS optimizations are done when all the radii of curvature and all separations between the optical surfaces were variable. Now the optimization algorithm has a possibility to change 17 construction parameters (9 radii of curvature and 8 separations between the optical surfaces). The merit functions of the initial optical system and the optimized optical systems are given in Table 15.2.

Table 15.2. The merit function values of the optimized Double Gauss objectives for different damping algorithms. The radii of curvature and the separations between optical surfaces are variable.

	Merit function
The initial optical system	21.387
The DLS optimization with the additive damping	1.839
The DLS optimization with the multiplicative damping	1.319

From Table 15.2 it can be seen that the DLS optimization with the multiplicative damping has found has found a better local minimum which is close to the best local minimum found by the evolution strategy ES REKO optimization. It is interesting to note that when only the radii of curvature were variable the DLS optimization with the additive damping has found optical system with a better merit function but when the radii of curvature and the separations between the optical surfaces are variable the DLS optimization with the multiplicative damping has found an optical system with a better merit function. This shows that when the number of variable construction parameters is small the DLS optimization with the additive damping is better in searching the optimization space. When the number of variable construction parameters is increased (almost doubled) the DLS optimization with the multiplicative damping is better in searching the optimization space. It is interesting to compare the optical system merit function values found by the DLS optimization when variable were only the radii of curvature and the radii of curvature and the separations between the optical surfaces. The number of variable construction parameters was in the former case 9 and in the latter case 17. As it can be expected, the DLS optimization with more variable construction parameters has found optical systems with a better merit function. The merit function for the best found optical system is reduced for approximately 2.3 times.

The last series of DLS optimizations are done when all construction parameters are variable (all radii of curvature, all separations between the optical surfaces and all glasses). The total number of variable construction parameters is 23 (9 radii of curvature, 8 separations between the optical surfaces and 6 glasses). The DLS optimization algorithm has a problem with the discrete variable parameters like glasses because it expects the variable parameters to be continuous. The radii of curvature and the separations between the optical surfaces are continuous parameters but the glasses are discrete ones (i.e. there is a finite number of glasses in the manufactures glass database). A variable parameter ought to be continuous because the partial derivatives of aberrations with respect to the variable construction parameters are calculated by adding and subtracting increments from the variable parameter and calculating aberrations. The discrete parameter must be transformed to the continuous parameter and the whole optimization must be done with this continuous parameter. The DLS optimization algorithm usually finds a good local minimum with this continuous parameter, but when these continuous parameters are transformed to the real glasses from the glass database, the merit function of this real optical system may be and usually is quite different from the merit function of the optical system with the variable glasses as continuous parameters. The conversion from the glass continuous parameter to the nearest real glass is done so that:
− the difference between the calculated index of refraction and the real glass index of refraction is minimum;
− the difference between the calculated Abbe number and the real glass Abbe number is minimum.

In the author's opinion the DLS optimization with the variable glasses ought to be used only when it is absolutely necessary. The optical designer can select good glasses from the glass database and run the DLS optimization with only radii of curvature and separations between the optical surfaces variable.

The merit functions of the initial optical system and the optimized optical systems are given in Table 15.3.

Table 15.3. The merit function values of the optimized Double Gauss objectives for different damping algorithms. The radii of curvature, the separations between optical surfaces and the glasses are variable.

	Merit function
The initial optical system	21.387
The DLS optimization with the additive damping	4.161
The DLS optimization with the multiplicative damping	3.125

From Table 15.3 it is clear that the DLS optimization has a problem in searching the optimization space and finding good optical systems when the glasses are variable. When the glasses are variable the DLS optimization has found the optical system with the merit function value that is worse than when the glasses are not variable (3.125 and 1.319). If the optimization results for the DLS optimization with the additive damping and with the multiplicative damping are compared then the DLS optimization with the multiplicative damping found optical systems with a better merit function. This is absolutely the same situation as when the radii of curvature and the separations between the optical surfaces are only variable.

15.4 ASSGA optimizations

The mathematical theory of the ASSGA optimization is given in Chapter 3 and the implementation of this optimization algorithm in the ADOS program is given in Chapter 9. The ASSGA optimizations for the selected Double Gauss objective are calculated for the following conditions:

- all radii of curvature are variable;
- all radii of curvature and all separations between the optical surfaces are variable;
- all radii of curvature, all separations between the optical surfaces and all glasses are variable.

In all ASSGA optimizations the following strategy is accepted. Each series of ASSGA optimizations is started with all default parameters. Two optimization parameter values: the population size and the starting values for genetic operator probabilities are independently varied in order to find the best combination.

Since the probabilistic events play an important role in the ASSGA optimization, it is necessary to run several optimizations to see a general behaviour of the optimization algorithm. The author decided to run five optimizations for each changed optimization parameter.

The first ASSGA optimizations of the Double Gauss objective are calculated when all radii of curvature are variable. There are nine radii of curvature that can be varied. As in the DLS optimizations there are enough variable construction parameters and the ASSGA optimization algorithm can search the optimization space successfully and find optical systems with a significantly better merit function than the initial optical system merit function. The merit functions of the optimized optical systems for different population sizes and starting values for the genetic operator probabilities are given in Table 15.4.

Table 15.4. The merit functions of the optimized Double Gauss objectives for different population sizes and different starting values for the genetic operator probabilities. Only radii of curvature are variable.

	Merit function						
	1st opt.	2nd opt.	3rd opt.	4th opt.	5th opt.	mean value	stand. deviat.
Merit function of the initial Double Gauss objective is 21.387							
$AC = UC = FM = BM = SM = 20$							
Population 300	24.218	30.685	60.094	54.132	43.395	42.505	15.162
Population 400	7.731	38.519	20.219	9.875	41.350	23.539	15.726
Population 500	36.122	41.861	4.470	29.382	42.659	30.899	15.705
$AC = UC = 30, FM = 20, BM = SM = 10$							
Population 400	10.718	30.488	14.427	40.010	50.565	29.242	16.843
Population 500	24.247	9.178	12.545	32.707	34.062	22.548	11.374
Population 600	26.421	21.125	26.774	8.052	17.660	20.006	7.691
Population 700	24.994	20.047	28.779	26.406	7.231	21.491	8.587
$AC = UC = 15, FM = 30, BM = SM = 20$							
Population 400	37.487	26.911	21.606	5.068	29.245	24.063	12.062
Population 500	5.559	11.235	34.151	15.757	13.003	16.061	10.780
Population 600	17.052	56.700	17.196	18.945	23.856	26.750	16.968

where:

AC is the average crossover;

UC is the uniform crossover;

FM is the floating-point number mutation;

BM is the big number creep;

SM is the small number creep.

From Table 15.4 it can be seen that the optimizations of the Double Gauss objective are calculated for the population sizes from 300 optical systems to 700 optical systems and for the following three sets of the starting values for the genetic operator probabilities:

− The first set is the default values for all genetic operator probabilities. The starting values for the genetic operator probabilities are the following: the average crossover, the uniform crossover, the floating point number mutation, the big number creep and the small number creep are equal to 20 ($AC = UC = FM = BM = SM = 20$).

− The second set is the set with larger values for the crossover probabilities than for the mutation probabilities. The starting values for the genetic operator probabilities are the following: the average crossover and the uniform crossover have the probability 30, the floating point number mutation has the probability 20, the big number creep and the small number creep have the probability 10 ($AC = UC = 30$, $FM = 20$, $BM = SM = 10$).

— The third set is the set with larger values for mutation probabilities than for the crossover probabilities. The starting values for the genetic operator probabilities are the following: the average crossover and the uniform crossover have the probability 15, the floating point number mutation has the probability 30, the big number creep and the small number creep have the probability 20 ($AC = UC = 30$, $FM = 20$, $BM = SM = 10$).

From Table 15.4 it can be seen that nine variable construction parameters are not enough for the ASSGA optimization to search the optimization space successfully. The ASSGA optimization has found a lot of different optical systems with the merit functions from 4.470 to 60.094. Only one optical system has a merit function similar to the optical system merit functions found by the DLS optimization and all other optical systems have greater merit functions. The best optical system with the smallest merit function (4.470) is found when the population size was 500 optical systems and the starting values for all genetic operator probabilities are equal. The optical systems with the smallest merit function mean value are found when the population size was 500 optical systems and the starting values for the mutation operators were greater than for the crossover operators.

The second set of the Double Gauss objective ASSGA optimizations is calculated when all radii of curvature and all separations between the optical surfaces are variable. The number of variable construction parameters is 17 (9 radii of curvature and 8 separations between the optical surfaces). The population size and the starting values for the genetic operator probabilities are independently varied. The genetic operator probabilities have three sets of values described on previous page. The merit functions of the optimized optical systems for different population sizes and starting values for the genetic operator probabilities are given in Table 15.5.

From Table 15.5 it can be seen that seventeen variable construction parameters are not enough for the ASSGA optimization to search the optimization space successfully. The ASSGA optimization has found a lot of different optical systems with the merit functions from 6.404 to 55.633. Two optical systems with the smallest merit functions (6.404 and 6.546) are found when the population sizes were 500 and 800 optical systems respectively. In the first case the starting values for the crossover operators were greater than for the mutation operators and in the second case the starting values for the mutation operators were greater than for the crossover operators. Both optical systems have approximately five times greater merit function than the merit function of the optical system found by the DLS optimization.

From the results presented in Table 15.5 very interesting phenomenon can be seen. The basic assumption in the genetic algorithms is that a larger number of variable construction parameters needs a larger population size. This is logical because of the implicit parallelism of genetic algorithms, the ASSGA optimization with a larger population size can search better optimization space. Here the situation is contrary to this basic assumption. When only the radii of curvature are variable, the Double Gauss objective has nine variable construction parameters and the optimum population size is 600 optical systems when the starting values for the crossover operators were greater than for the mutation operators. When the radii of curvature and the separations between the optical surfaces are variable, the Double Gauss objective has seventeen variable construction parameters (the number of variable parameters is nearly doubled) and the optimum population size is 400 optical systems when the starting values for the crossover operators were greater than for the mutation operators. The criterion for the selection of the optimum population size is the minimum merit function mean value.

Table 15.5. The merit functions of the optimized Double Gauss objectives for different population sizes and the different starting values for the genetic operator probabilities. The radii of curvature and the separations between the optical surfaces are variable.

	Merit function						
	1st opt.	2nd opt.	3rd opt.	4th opt.	5th opt.	mean value	stand. deviat.
Merit function of the initial Double Gauss objective is 21.387							
$AC = UC = FM = BM = SM = 20$							
Population 500	11.872	50.698	41.044	45.665	14.069	32.670	18.320
Population 600	8.673	16.669	18.494	13.846	14.945	14.525	3.716
Population 700	17.234	20.424	44.987	15.092	18.252	23.198	12.331
$AC = UC = 30, FM = 20, BM = SM = 10$							
Population 300	49.155	12.326	51.879	8.679	52.701	34.948	22.391
Population 400	14.824	7.718	14.032	11.771	21.520	13.973	5.041
Population 500	6.404	15.477	26.886	25.385	41.637	23.158	13.229
Population 600	30.993	29.962	20.955	52.162	29.854	32.785	11.567
$AC = UC = 15, FM = 30, BM = SM = 20$							
Population 400	39.598	43.261	55.633	21.371	24.577	36.888	14.068
Population 500	37.534	10.275	34.825	18.145	13.557	22.867	12.507
Population 600	12.612	17.667	16.978	24.268	34.402	21.185	8.482
Population 700	15.492	9.064	13.612	13.689	12.059	12.783	2.408
Population 800	15.592	6.546	15.179	10.505	12.821	12.139	3.730
Population 900	9.988	12.294	21.323	19.732	9.817	14.631	5.500

where:

AC is the average crossover;

UC is the uniform crossover;

FM is the floating-point number mutation;

BM is the big number creep;

SM is the small number creep.

The third set of the ASSGA optimizations of the Double Gauss objective is calculated when all radii of curvature, all separations between the optical surfaces and all glasses are variable. The number of variable construction parameters is 23 (9 radii of curvature, 8 separations between the optical surfaces and 6 glasses). The ASSGA optimization algorithm does not require all variables to be continuous, so the ASSGA optical system can work directly with glasses from the glass database. In fact the ASSGA optimization algorithm randomly selects a glass from the glass database. The ASSGA optimization algorithm has enough variable construction parameters to search the optimization space successfully and find the optical system with the best merit function. The ASSGA optimization has found much better optical systems with a smaller merit function when the glasses are variable than when the glasses are constant (not variable). Other optimization algorithms usually have found better optical systems when only the radii of curvature and the separations between the optical surfaces are variable. The following optimization parameters are independently varied: the population size and the starting values for the genetic operator probabilities. The genetic operator probabilities have three sets of values previously described. All other optimization parameters have default values. The merit functions of the optimized optical systems for different population sizes and starting values for the genetic operator probabilities are given in Table 15.6.

Table 15.6. The merit functions of the optimized Double Gauss objectives for different population sizes and different starting values for the genetic operator probabilities. The radii of curvature, the separations between the optical surfaces and the glasses are variable.

	Merit function						
	1st opt.	2nd opt.	3rd opt.	4th opt.	5th opt.	mean value	stand. deviat.
Merit function of the initial Double Gauss objective is 21.387							
$AC = UC = FM = BM = SM = 20$							
Population 500	5.250	19.287	11.518	23.714	22.844	16.523	7.928
Population 600	7.657	17.161	12.071	7.858	11.862	11.322	3.885
Population 700	8.409	9.372	11.369	11.966	11.342	10.491	1.523
Population 800	14.794	14.030	15.015	2.990	11.754	11.717	5.045
$AC = UC = 30, FM = 20, BM = SM = 10$							
Population 500	19.396	5.670	20.340	20.864	17.979	16.850	6.345
Population 600	11.727	10.221	21.309	21.304	5.046	13.921	7.183
Population 700	15.741	21.371	10.474	15.947	12.810	15.269	4.091
$AC = UC = 15, FM = 30, BM = SM = 20$							
Population 400	23.800	14.485	27.466	10.617	29.353	21.144	8.209
Population 500	5.147	2.093	4.587	2.731	2.799	3.471	1.318
Population 600	19.582	19.582	2.054	4.207	3.126	9.710	9.044
Population 700	10.192	17.540	8.485	22.630	18.804	15.530	5.985

where:

AC is the average crossover;

UC is the uniform crossover;

FM is the floating-point number mutation;

BM is the big number creep;

SM is the small number creep.

The ASSGA optimization has found a lot of different optical systems with the merit functions from 2.054 to 27.466. The best optical systems are found when the starting values for the mutation operators were greater than for the crossover operators. The best optical system with the minimum merit function is found when the population size was 600 optical systems. The optical systems with the smallest merit function mean value are found when the population size was 500 optical systems. Three out of five optical systems have merit function smaller than three.

From the results presented in Table 15.6 the phenomenon described when the radii of curvature and the separations between the optical surfaces were variable has happened also when all construction parameters were variable. When the radii of curvature are only variable, the Double Gauss objective has nine variable construction parameters and the optimum population size is 500 optical systems when the starting values for the mutation operators were greater than for the crossover operators. When the radii of curvature and the separations between the optical surfaces are variable, the Double Gauss objective has seventeen variable construction parameters (the number of variable parameters is nearly doubled) and the optimum population size is 800 optical systems when the starting values for the mutation operators were greater than for the crossover operators. This is standard and expected behaviour of the ASSGA optimization but the problem raises when all construction parameters are variable. When the radii of curvature, the separations between the optical surfaces and the glasses are variable, the Double Gauss objective has 23 variable construction parameters and the optimum population size is 500 optical systems when the starting values for the mutation operators were greater than for the crossover operators. The criterion for the selection of the optimum population size is the minimum merit function mean value.

It is interesting to compare the best merit function value found by the ASSGA optimization when the radii of curvature the separations between the optical surfaces were only variable and the radii of curvature, the separations between the optical surfaces and the glasses were variable. The number of variable construction parameters is increased from 17 to 23 and the best merit function value is reduced for more than three times from 6.404 to 2.054. If compared with the DLS optimization the ASSGA optimization has found a better optical system with the approximately 50% smaller merit function. The ASSGA optimization has found the optical system with the merit function 2.054 and the DLS optimization has found the optical system with the merit function 3.125.

15.5   Evolution strategies optimizations

The evolution strategies are global optimization algorithms based on the theory of evolution and natural selection developed for the optimization of complex technical systems. In this monograph the mathematical theory for four different evolution strategies is given in Chapter 4 and their implementation in the ADOS

program is given in Chapters 10 – 12. Here will be presented the optimizations of the selected Double Gauss objective with the following evolution strategies:
- the two membered evolution strategy ES EVOL;
- the multimembered evolution strategy ES GRUP;
- the multimembered evolution strategy ES REKO;
- the multimembered evolution strategy ES KORR.

15.5.1   Two membered evolution strategy ES EVOL optimizations

The two-membered evolution strategy ES EVOL represents the most simple evolution strategy. The ES EVOL optimizations for the selected Double Gauss objective are calculated for the following conditions:
- all radii of curvature are variable;
- all radii of curvature and all separations between the optical surfaces are variable;
- all radii of curvature, all separations between the optical surfaces and all glasses are variable.

The two membered evolution strategy ES EVOL has a small number of optimization parameters and all parameters are kept at default values.

The first ES EVOL optimizations of the Double Gauss objective are calculated when all radii of curvature are variable. There are nine radii of curvature that can be varied. As in the DLS optimizations there are enough variable construction parameters and the ES EVOL optimization algorithm can search optimization space successfully and find the optical systems with a significantly better merit function than the initial optical system merit function. The merit functions of the optimized optical systems are given in Table 15.7.

Table 15.7. The merit functions of the optimized Double Gauss objectives. Only radii of curvature are variable.

	Merit function
The 1st optimization	3.552
The 2nd optimization	3.783
The 3rd optimization	3.706
The 4th optimization	3.789
The 5th optimization	3.748
The merit function mean value	3.716
The merit function standard deviation	0.097

The ES EVOL optimizations have found the optical systems with very similar merit functions from 3.552 to 3.789. The merit function mean value is 3.716 and the merit function standard deviation is 0.097. These results show that all ES EVOL optimizations have found almost the same local minimum. It is interesting to compare optical system merit functions found by various optimization algorithms.

The ES EVOL optimization has found the optical system with a better merit function than the ASSGA optimization and with a worse merit function than the DLS optimization.

The second group of the ES EVOL optimizations of the Double Gauss objective is calculated when all radii of curvature and all separations between the optical surfaces are variable. The number of variable construction parameters is 17 (9 radii of curvature and 8 separations between the optical surfaces). The merit functions of the optimized optical systems are given in Table 15.8.

Table 15.8. The merit functions of the optimized Double Gauss objectives. The radii of curvature and the separations between the optical surfaces are variable.

	Merit function
The 1st optimization	3.361
The 2nd optimization	3.646
The 3rd optimization	3.682
The 4th optimization	2.779
The 5th optimization	3.228
The merit function mean value	3.339
The merit function standard deviation	0.367

The ES EVOL optimization has found different optical systems with the merit functions from 2.779 to 3.682. It is interesting to note that the number of variable construction parameters is nearly doubled from nine to seventeen and the optical system merit function is reduced for only approximately 28% from 3.552 to 2.779. If the optical system merit functions found by various optimization algorithms are compared then the ES EVOL optimization has found the optical system with a better merit function than the ASSGA optimization and with a worse merit function than the DLS optimization.

The third group of the ES EVOL optimizations of the Double Gauss objective is calculated when all radii of curvature, all separations between the optical surfaces and all glasses are variable. The number of variable construction parameters is 23 (9 radii of curvature, 8 separations between the optical surfaces and 6 glasses). The merit functions of the optimized optical systems are given in Table 15.9.

Table 15.9. The merit functions of the optimized Double Gauss objectives. The radii of curvature, the separations between the optical surfaces and the glasses are variable.

	Merit function
The 1st optimization	2.952
The 2nd optimization	2.915
The 3rd optimization	3.046
The 4th optimization	2.015
The 5th optimization	1.396
The merit function mean value	2.465
The merit function standard deviation	0.728

The ES EVOL optimization has found different optical systems with the merit functions from 1.396 to 3.046. The ES EVOL optimization has found the best optical system with the minimum merit function when all optimization algorithms are considered. It is interesting to compare the best merit function value found by the ES EVOL optimization when the radii of curvature the separations between the optical surfaces were only variable and the radii of curvature, the separations between the optical surfaces and the glasses were variable. The number of variable construction parameters is increased from 17 to 23 and the best merit function value is reduced approximately two times from 2.779 to 1.396.

15.5.2 Multimembered evolution strategy ES GRUP optimizations

The multimembered evolution strategy ES GRUP represents the further development of ideas from the two membered evolution strategy ES EVOL. The ES GRUP optimizations for the selected Petzval objective are calculated for the following conditions:

− all radii of curvature are variable;
− all radii of curvature and all separations between the optical surfaces are variable;
− all radii of curvature, all separations between the optical surfaces and all glasses are variable.

The multimembered evolution strategy ES GRUP has a small number of optimization parameters and all parameters are kept at default values.

The first ES GRUP optimizations of the Double Gauss objective are calculated when all radii of curvature are variable. There are nine radii of curvature that can be varied. As in the DLS and the ES EVOL optimizations there are enough variable construction parameters and the ES GRUP optimization algorithm can search optimization space successfully and find optical systems with a significantly better merit function than the initial optical system merit function. The merit functions of the optimized optical systems are given in Table 15.10.

Table 15.10. The merit functions of the optimized Double Gauss objectives. Only radii of curvature are variable.

	Merit function
The 1st optimization	4.504
The 2nd optimization	4.843
The 3rd optimization	4.746
The 4th optimization	4.718
The 5th optimization	5.098
The merit function mean value	4.782
The merit function standard deviation	0.216

The ES GRUP optimization has found different optical systems with the merit functions from 4.504 to 5.098. The merit function mean value is 4.782 and the merit function standard deviation is 0.216. The ES GRUP optimization that represents the further improvement of the ES EVOL optimization has found optical systems with worse merit functions than the ES EVOL optimization. The best merit function values are 3.552 for the ES EVOL optimization and 4.504 for the ES GRUP optimization. The merit function mean values are 3.716 for the ES EVOL optimization and 4.782 for the ES GRUP optimization. It is interesting to compare the ES GRUP optimization with the other optimization algorithms (the DLS optimization and the ASSGA optimization). If only the optical systems with the minimum merit function are considered then the ES GRUP optimization has found the optical system with the largest merit function. It has a little larger merit function than the optical system found by the ASSGA optimization and it has significantly larger merit function than the optical systems found by the DLS optimization and the ES EVOL optimization. In order to completely analyze the optimization behaviour it is necessary to consider the best merit function, the merit function mean value and the merit function standard deviation. The ES GRUP optimization has approximately from three to six times smaller merit function mean value then the ASSGA optimization.

The second group of the ES GRUP optimizations of the Double Gauss objective is calculated when all radii of curvature and all separations between the optical surfaces are variable. The number of variable construction parameters is 17 (9 radii of curvature and 8 separations between the optical surfaces). The merit functions of the optimized optical systems are given in Table 15.11.

Table 15.11. The merit functions of the optimized Double Gauss objectives. The radii of curvature and the separations between the optical surfaces are variable.

	Merit function
The 1st optimization	4.351
The 2nd optimization	4.670
The 3rd optimization	5.186
The 4th optimization	3.694
The 5th optimization	5.111
The merit function mean value	4.602
The merit function standard deviation	0.611

The ES GRUP optimization has found the optical system with significantly different merit functions from 3.694 to 5.186. This means that the optimization algorithm has found quite different local minimums. It is interesting to note that the number of variable construction parameters is nearly doubled from nine to seventeen and the best optical system merit function is reduced for only approximately 22% from 4.504 to 3.694. The merit function mean value is reduced for only approximately 4% from 4.782 to 4.602. The ES GRUP optimization that represents the further improvement of the ES EVOL optimization has found optical systems with worse merit functions than the ES EVOL optimization. The best merit function values are 2.779 for the ES EVOL optimization and 3.694 for the ES GRUP optimization. The merit function mean values are 3.339 for the ES EVOL optimization and 4.602 for the ES GRUP optimization. It is interesting to compare the ES GRUP optimization with the other optimization algorithms (the DLS optimization and the ASSGA optimization). The ES GRUP optimization is similar to the ES EVOL optimization and it has found the optical systems with a better merit function than the ASSGA optimization and with a worse merit function than the DLS optimization.

The third group of the ES GRUP optimizations of the Double Gauss objective is calculated when all radii of curvature, all separations between the optical surfaces and all glasses are variable. The number of variable construction parameters is 23 (9 radii of curvature, 8 separations between the optical surfaces and 6 glasses). The merit functions of the optimized optical systems are given in Table 15.12.

Table 15.12. The merit functions of the optimized Double Gauss objectives. The radii of curvature, the separations between the optical surfaces and the glasses are variable.

	Merit function
The 1st optimization	2.743
The 2nd optimization	2.398
The 3rd optimization	3.570
The 4th optimization	3.127
The 5th optimization	3.728
The merit function mean value	3.113
The merit function standard deviation	0.556

The ES GRUP optimization has found the optical system with significantly different merit functions from 2.398 to 3.728. This means that the optimization algorithm has found quite different local minimums. The comparison of the optical systems found by the ES GRUP optimization when the radii of curvature and the separations between the optical surfaces were only variables and when all construction parameters were variable shows that the ES GRUP optimization with all variable construction parameters has found better optical systems with approximately 49% smaller merit functions. As in all preceding optimizations the ES GRUP optimization although a further improvement of the ES EVOL optimization has found optical systems with worse merit functions than the ES EVOL optimization.

It is interesting to compare the ES GRUP optimization with the other optimization algorithms (the DLS optimization and the ASSGA optimization). If only the optical systems with the minimum merit function are considered then the ES GRUP optimization has found the optical system with a merit function larger than the ASSGA optimization and smaller than the DLS optimization. In order to completely analyze the optimization behaviour it is necessary to consider the best merit function, the merit function mean value and the merit function standard deviation. The ES GRUP optimization has approximately from 10% to three times smaller merit function mean value then the ASSGA optimization.

15.5.3   Multimembered evolution strategy ES REKO optimizations

The multimembered evolution strategy ES REKO represents the further development of ideas from the two membered evolution strategy ES EVOL and the multimebered evolution strategy ES GRUP. The ES REKO optimizations for the selected Double Gauss objective are calculated for the following conditions:
- all radii of curvature are variable;
- all radii of curvature and all separations between the optical surfaces are variable;
- all radii of curvature, all separations between the optical surfaces and all glasses are variable.

The multimembered evolution strategy ES REKO has the same set of optimization parameters as the ES REKO optimization and all parameters are kept at default values.

The first ES REKO optimizations of the Double Gauss objective are calculated when all radii of curvature are variable. There are nine radii of curvature that can be varied. As in all preceding optimizations there are enough variable construction parameters and the ES REKO optimization algorithm can search optimization space successfully and find optical systems with a significantly better merit function than the initial optical system merit function. The merit functions of the optimized optical systems are given in Table 15.13.

Table 15.13. The merit functions of the optimized Petzval objectives. Only radii of curvature are variable.

	Merit function
The 1st optimization	2.869
The 2nd optimization	2.897
The 3rd optimization	3.039
The 4th optimization	2.880
The 5th optimization	2.866
The merit function mean value	2.910
The merit function standard deviation	0.073

The ES REKO optimizations have found the optical systems with similar merit functions from 2.869 to 3.039. The merit function mean value is 2.910 and the merit function standard deviation is 0.073. The ES REKO optimization has found the optical systems which are among the best optical systems found by any optimization algorithm. The smallest difference in the merit function values is with the DLS optimization and the largest difference in the merit function values is with the ASSGA optimization and the ES GRUP optimization.

The second group of the ES REKO optimizations of the Double Gauss objective is calculated when all radii of curvature and all separations between the optical surfaces are variable. The number of variable construction parameters is 17 (9 radii of curvature and 8 separations between the optical surfaces). The merit functions of the optimized optical systems are given in Table 15.14.

Table 15.14. The merit functions of the optimized Double Gauss objectives. The radii of curvature and the separations between the optical surfaces are variable.

	Merit function
The 1st optimization	1.347
The 2nd optimization	1.327
The 3rd optimization	1.409
The 4th optimization	1.246
The 5th optimization	1.584
The merit function mean value	1.383
The merit function standard deviation	0.127

The ES REKO optimizations have found the optical systems with similar merit functions from 1.246 to 1.584. The merit function mean value is 1.383 and the merit function standard deviation is 0.127. When the number of variable construction parameters is nearly doubled from nine to seventeen the ES REKO optimization has found, as expected, the optical systems with approximately 2.1 times smaller merit functions. The merit function mean value when only radii of curvature are variable is 2.910 and the merit function mean value when the radii of curvature and the separations between the optical surfaces are variable is 1.383. The ES REKO optimization has found the best optical systems with better merit functions than any optimization algorithm. The smallest difference in the merit function values is with the DLS optimization and the largest difference in the merit function values is with the ASSGA optimization.

The third group of the ES REKO optimizations of the Double Gauss objective is calculated when all radii of curvature, all separations between the optical surfaces and all glasses are variable. The number of variable construction parameters is 23 (9 radii of curvature, 8 separations between the optical surfaces and 6 glasses). The ES REKO optimization and the ES KORR are the only evolution strategies that have the problem with variable glasses in the Double Gauss objective optimization. The evolution strategies are the optimization algorithms designed to work with continuous parameters. The variable glasses are discrete parameters that are selected from the database with the finite number of different glasses. All evolution strategies have the problems with the variable glasses in the Cooke triplet optimizations and the Petzval objective optimizations. A detailed description of this problem is given in Chapters 13 and 14 and it will not be given here. The merit functions of the optimized optical systems are given in Table 15.15.

From Table 15.15 it can be seen that the ES REKO optimization has found different optical systems with the merit functions from 2.041 to 3.608. The merit function mean value is 3.046 and the merit function standard deviation is 0.592. The comparison of the optical systems found by the ES REKO optimization when the radii of curvature and the separations between the optical surfaces were only variables and when all construction parameters were variable shows that the ES

REKO optimization with all variable construction parameters has found the optical systems with significantly worse merit functions. If the best merit functions are compared then the merit function increment is approximately 64%. If the merit function mean values are compared then the merit function increment is approximately 2.2 times.

Table 15.15. The merit functions of the optimized Petzval objectives. The radii of curvature, the separations between the optical surfaces and the glasses are variable.

	Merit function
The 1st optimization	2.041
The 2nd optimization	3.608
The 3rd optimization	3.253
The 4th optimization	3.218
The 5th optimization	3.112
The merit function mean value	3.046
The merit function standard deviation	0.592

If the ES REKO optimization is compared with the other optimizations then only the ES EVOL optimization has found better optical systems and the DLS optimization, the ASSGA optimization and the ES GRUP optimization have found worse optical systems.

15.5.4   Multimembered evolution strategy ES KORR optimizations

The multimembered evolution strategy ES KORR is the most developed and the most complex evolution strategy. The ES KORR optimizations for the selected Double Gauss objective are calculated for the following conditions:

– all radii of curvature are variable;

– all radii of curvature and all separations between the optical surfaces are variable;

– all radii of curvature, all separations between the optical surfaces and all glasses are variable.

All optimization parameters except one, in the multimembered evolution strategy ES KORR are kept at default values. The only variable optimization parameter is the rotation of the mutation hyperellipsoid. The basic idea about the rotation of mutation hyperellipsoid and its influence on variable construction parameters is given in Section 13.5.4.

The first ES KORR optimizations of the Double Gauss objective are calculated when all radii of curvature are variable. There are nine radii of curvature that can be varied. As in all preceding optimizations there are enough variable construction parameters and the ES KORR optimization algorithm can search optimization space successfully and find optical systems with a significantly better merit function than the initial optical system merit function. The merit functions of the optimized optical systems are given in Table 15.16.

Table 15.16. The merit functions of the optimized Double Gauss objectives. Only radii of curvature are variable.

	Merit function	
	The mutation hyperellipsoid cannot rotate	The mutation hyperellipsoid can rotate
The 1st optimization	2.850	7.720
The 2nd optimization	2.854	7.629
The 3rd optimization	2.850	6.452
The 4th optimization	2.857	3.980
The 5th optimization	2.851	6.274
The merit function mean value	2.852	6.321
The merit function standard deviation	0.003	1.424

From Table 15.16 it can be seen that the ES KORR have found quite different optical systems. Usually the possibility of mutation hyperellipsoid rotation enables a better search of the optimization space. But in the Double Gauss optimization it is not the case. There the rotation of the mutation hyperellipsoid prevents the optimization algorithm to search optimization space successfully. The optical systems, found by the ES KORR optimiza-tion when the mutation hyperellipsoid had the possibility of rotation, have merit functions from 3.980 to 7.629 with the mean value 6.321 and the standard deviation 1.421. This is more than two times worse than when the mutation hyperellipsoid cannot rotate.

The optical systems, found by the ES KORR optimization when the mutation hyperellipsoid could not rotate, have the merit functions from 2.850 to 2.857. All ES KORR optimizations have found optical systems with almost the same merit function. It is interesting to note that the ES KORR optimization, when the mutation hyperellipsoid cannot rotate, have found the best optical systems in all optimization algorithms. All five optical systems found by the ES KORR optimization have a better merit function than the next best optical system found by the ES REKO optimization. Also the ES KORR optimization, when the mutation hyperellipsoid can rotate, has found the worst optical systems in all evolution strategies.

The second group of the ES KORR optimizations of the Double Gauss objective is calculated when all radii of curvature and all separations between the optical surfaces are variable. The number of variable construction parameters is 17 (9 radii of curvature and 8 separations between the optical surfaces). The merit functions of the optimized optical systems are given in Table 15.17.

From Table 15.17 it can be seen that the ES KORR has found quite different optical systems. The situation is the same when only radii of curvature were variable. Here the rotation of the mutation hyperellipsoid also prevents the optimization algorithm to search optimization space successfully. The optical systems, found by the ES KORR optimization when the mutation hyperellipsoid had the possibility of rotation, have the merit functions from 4.772 to 6.360 with the mean value 5.523 and the standard deviation 0.576. This is approximately three times worse than when the mutation hyperellipsoid cannot rotate. The optical systems, found by the ES KORR optimization when the mutation hyperellipsoid could not rotate, have the merit functions from 1.555 to 2.341 with the mean value 1.834 and the standard deviation 0.324. It is interesting to note that the ES KORR optimization, when the mutation hyperellipsoid cannot rotate, has found the optical systems which are among the best optical systems found by any optimization algorithm. Also the ES KORR optimization, when the mutation hyperellipsoid can rotate, has found the worst optical systems in all evolution strategies. The comparison of the optical systems found by the ES KORR when the radii of curvature are variable and the ES KORR when the radii of curvature and the separations between the optical surfaces are variable shows that the variable construction parameters are nearly doubled and the optical system merit function mean value is reduced for approximately 55% from 2.852 to 1.834.

Table 15.17. The merit functions of the optimized Double Gauss objectives. The radii of curvature and the separations between the optical surfaces are variable.

	Merit function	
	The mutation hyperellipsoid cannot rotate	The mutation hyperellipsoid can rotate
The 1st optimization	1.953	5.471
The 2nd optimization	1.589	6.360
The 3rd optimization	1.733	5.333
The 4th optimization	1.555	4.772
The 5th optimization	2.341	5.677
The merit function mean value	1.834	5.523
The merit function standard deviation	0.324	0.576

The third group of the ES KORR optimizations of the Double Gauss objectives is calculated when all radii of curvature, all separations between the optical surfaces and all glasses are variable. The number of variable construction parameters is 23 (9 radii of curvature, 8 separations between the optical surfaces and 6 glasses). The problem with variable glasses is the same as in ES REKO optimizations and it is described in Section 15.5.3. The consequence of this problem is a very slow convergence to the minimum, which is not the best possible local minimum and a large number of executed mutations with a small number of successful mutations. The merit functions of the optimized optical systems are given in Table 15.18.

From Table 15.18 it can be seen that the ES KORR optimization in both cases has found the optical systems with the similar merit functions. The ES KORR optimization with the mutation hyperellipsoid rotation has found the optical system with the best merit function 1.568 and the merit function mean value 2.535. The ES KORR optimization without the mutation hyperellipsoid rotation has found the optical system with the best merit function 1.661 and the merit function mean value 3.605. Until now the ES KORR without the mutation hyperellipsoid rotation has found optical systems with smaller merit functions. This is only the case with the ES KORR optimizations of the Double Gauss optimizations when the optical systems with the smaller merit functions are found when the mutation hyperellipsoid can rotate. The comparison of the optical systems found by the ES KORR optimization when the radii of curvature and the separations between the optical surfaces were only variables and when all construction parameters were variable shows that the ES KORR optimization with all variable construction parameters has found the optical systems with the approximately 38% worse merit function mean value. This increment in the merit function value can be explained by bad influence of glass variation because the glasses are discrete variables and evolution strategies are designed for continuous variables. The ES KORR optimization has found the optical system with a better merit function than all optimization algorithms except the ES EVOL optimization.

Table 15.18. The merit functions of the optimized Double Gauss objectives. The radii of curvature, the separations between the optical surfaces and the glasses are variable.

	Merit function	
	The mutation hyperellipsoid can rotate	The mutation hyperellipsoid cannot rotate
The 1st optimization	3.067	3.905
The 2nd optimization	4.474	4.881
The 3rd optimization	1.666	4.757
The 4th optimization	1.899	1.661
The 5th optimization	1.568	2.822
The merit function mean value	2.535	3.605
The merit function standard deviation	1.239	1.364

15.6 Comparison of the Double Gauss objective designs by various optimization methods

All optimization methods: the DLS optimization, the ASSGA optimization, the two membered evolution strategy ES EVOL optimization, the multimembered evolution strategies ES GRUP, ES REKO and ES KORR optimizations have calculated various Double Gauss objectives with the following variable construction parameters:

— all radii of curvature;

— all radii of curvature and all separations between the optical surfaces;

— all radii of curvature and all separations between the optical surfaces and all glasses.

The first group of the Double Gauss objective optimizations is calculated when all radii of curvature are variable. There are nine radii of curvature that can be varied. The optical systems with the best merit function found by any optimization algorithm, the merit function mean value and the merit function standard deviation calculated for any stochastic optimization algorithm are given in Table 15.19.

Table 15.19. The best merit function value, the merit function mean value and the merit function standard deviation for all optimization algorithms. Only radii of curvature are variable.

	The best merit function value	The merit function mean value	The merit function standard deviation
Initial optical system	21.387	-	-
The DLS optimization	3.044	-	-
The ASSGA optimization	4.470	30.899	15.705
The ES EVOL optimization	3.552	3.716	0.097
The ES GRUP optimization	4.504	4.782	0.216
The ES REKO optimization	2.866	2.910	0.073
The ES KORR optimization	2.850	2.852	0.003

From Table 15.19 it can be seen that all optical systems optimized by various optimization methods represent significant improvement to the starting optical system. All optimized optical systems have the merit functions from 4.75 to 7.50 times smaller than the starting optical system. The optical system with the largest merit function is found by the ASSGA optimization and the ES GRUP optimization. From Table 15.19 it can be seen that the evolution strategies have found quite different optical systems with the merit functions from 4.504 to 2.850. This difference in the merit function values can be explained by the genetic operators used in the optimization. The ES EVOL and the ES GRUP optimization use only one genetic operator – mutation and this genetic operator does not provide to the optimization algorithm the possibility to search the optimization space successfully. The result is that the optimization algorithm finds the local minimums which are sub optimal minimums and surely not the global minimum. The ES REKO and the ES KORR optimization use two types of genetic operators: the mutation operators and the recombination operators. This recombination operators are responsible for searching the optimization space successfully and finding the best local minimums that can be the global minimum. The ES REKO and the ES KORR optimization have found optical systems with very similar merit functions. Among the evolution strategies the best performing strategy is the ES KORR which:

– has found the optical system with the minimum merit function;

– has the minimum merit function mean value;

– has the minimum merit function standard deviation value.

The second group of the Double Gauss objective optimizations is calculated when all radii of curvature and all separations between the optical surfaces are variable. The number of variable construction parameters is 17 (9 radii of curvature and 8 separations between the optical surfaces). The optical systems with the best merit function found by any optimization algorithm, the merit function mean value and the merit function standard deviation calculated for any stochastic optimization algorithm are given in Table 15.20.

Table 15.20. The best merit function value, the merit function mean value and the merit function standard deviation for all optimization algorithms. The radii of curvature and the separations between the optical surfaces are variable.

	The best merit function value	The merit function mean value	The merit function standard deviation
Initial optical system	21.387	-	-
The DLS optimization	1.319	-	-
The ASSGA optimization	6.404	23.158	13.229
The ES EVOL optimization	2.779	3.339	0.367
The ES GRUP optimization	3.694	4.602	0.611
The ES REKO optimization	1.246	1.383	0.127
The ES KORR optimization	1.555	1.834	0.324

From Table 15.20 it can be seen that all optical systems optimized by various optimization methods represent a significant improvement to the starting optical system. All optimized optical systems have the merit functions from 3.34 to 17.16 times smaller than the starting optical system. The optical system with the largest merit function is found by the ASSGA optimization and the ES GRUP optimization. The evolution strategies have found quite different optical systems with the merit functions from 3.694 to 1.246. The reason for this difference in the merit function values is the same as in the optimizations when only radii of curvature are variable. The most important factor in all evolution strategies optimizations is the genetic operators. The ES EVOL and the ES GRUP optimization use only one genetic operator – mutation and this genetic operator does not provide to the optimization algorithm the possibility to search the optimization space successfully. The result is that the optimization algorithm finds the local minimums which are sub-optimal minimums and surely not the global minimum. The ES REKO and the ES KORR optimization use two types of genetic operators: the mutation operators and the recombination operators. These recombination operators, by combining good properties of the parents, are responsible for searching the optimization space successfully and finding the best local minimums that can be the global minimum. The ES REKO and the ES KORR optimization have found optical systems with similar merit functions. Among the evolution strategies the best performing strategy is the ES REKO which:

– has found the optical system with the minimum merit function;
– has the minimum merit function mean value;
– has the minimum merit function standard deviation value.

It is interesting to note that all optimization algorithms except the ASSGA optimization have found optical systems with a smaller merit function when the radii of curvature and the separations between the optical surfaces were variable than when only the radii of curvature were variable.

The third group of the Double Gauss objective optimizations is calculated when all radii of curvature, all separations between the optical surfaces and all glasses are variable. The number of variable construction parameters is 23 (9 radii of curvature,

8 separations between the optical surfaces and 6 glasses). The DLS optimization and all evolution strategies (the ES EVOL, the ES GRUP, the ES REKO and the ES KORR) expect that all variable construction parameters are continuous variables. The problem with the variable glasses is that they are not continuous variables but there are discrete variables that are chosen from the glass database. Each optimization algorithm overcomes this problem in a specific way. The ASSGA optimization does not have this problem. It can work with continuous and discrete variable construction parameters. The optical systems with the best merit function found by any optimization algorithm, the merit function mean value and the merit function standard deviation calculated by any optimization algorithm are given in Table 15.21.

Table 15.21. The best merit function value, the merit function mean value and the merit function standard deviation for all optimization algorithms. The radii of curvature, the separations between the optical surfaces and the glasses are variable.

	The best merit function value	The merit function mean value	The merit function standard deviation
Initial optical system	21.387	-	-
The DLS optimization	3.125	-	-
The ASSGA optimization	2.054	9.710	9.044
The ES EVOL optimization	1.396	2.465	0.728
The ES GRUP optimization	2.398	3.113	0.556
The ES REKO optimization	2.041	3.046	0.592
The ES KORR optimization	1.568	2.535	1.239

From Table 15.21 it can be seen that all optical systems optimized by various optimization methods represent significant improvement to the starting optical system. All optimized optical systems have the merit functions from 6.84 to 15.32 times smaller than the starting optical system. The most important factor in these optimizations is the way of dealing with the variable glasses which are discrete variables. The DLS optimization has the greatest difficulties with variable glasses so it has found the optical system with the worst merit function. The ASSGA optimization, which is the only optimization algorithm that can work with discrete variables, has found the optical system with good but not the best merit function. The evolution strategies: the ES EVOL, the ES REKO and the ES KORR have found better optical systems than the ASSGA optimization. The best optimization algorithm is the ES EVOL optimization which:

- has found the optical system with the minimum merit function;
- has the minimum merit function mean value.

The ES EVOL is the simplest implementation of the evolution strategies and in this case the most robust implementation with the smallest bad influence from variable glasses as a discrete variable.

Chapter 16

Summary and outlook

As it can be seen from the historical overview in Chapter 1, the development of optical design methods is firmly connected to the development of computers. Today a single most influential factor in the development of optical design procedures is availability of inexpensive and powerful personal computers. First optical designers were satisfied only to design an optical system because the computer speed was barely enough for this task. As soon as the computer speed was increased the desire for the optimization of optical system was born. First there were classical optimization methods optimization methods based on famous mathematical (numerical) methods. These optimization methods were very good and solved many problems. But one problem remained with these optimization methods that cannot be solved. These optimization methods requested a good knowledge of the optical system being optimized from the optical designer. For the classical optical systems like objectives this was not a great difficulty because they were very well researched. But for new types of optical systems or very specific optical systems this was a great problem because of no prior knowledge available. The knowledge of the optical system was used for selecting good starting points because the optimization methods were able to find only the local minimum closest to the starting point. There is no guarantee that the found local minimum is in fact the global minimum. The optical designer usually has to start optimization from several different starting points and to compare the optimization results. With the introduction of modern optimization methods like the genetic algorithms and the evolution strategies the problem of selecting good starting points is diminished. The evolutionary algorithms (the genetic algorithms and the evolution strategies) work with the population of randomly chosen starting systems so they have a possibility to escape from the local minimums that are not optimum solutions. Today very powerful optimization methods both classical and evolutionary are present and they allow design and optimization of almost any optical system.

But, anyone who has designed an optical system can tell that the design and the optimization of the optical system is not a whole story. Nobody can make a perfect optical system without manufacturing errors. The role of the optical designer is to analyze the influence of these manufacturing errors on the performance of optical systems. This is often called tolerancing the optical system. Today this is a great new field for research. It is important to find the optimization methods that will produce the optimal tolerances for optical systems. It will be very good if the optimization of optical systems and the tolerance optimization can be put together to produce an optical system with the required image quality and optimal tolerances for production.

The process of designing an optical system is partly art, partly science and a lot of patience. The optical designer who finds pleasure in the act of designing may well succeed, while one who loses interest early settles on a lower quality optical design. The optical designers of 50 and more years ago had an endurance that remains unmatched. This was a small community of dedicated professionals and they worked with classical optical systems. Today the situation has changed. There is a constant demand for more and more optical, optoelectronical and fiber optical systems. Various complex technical systems have the optical components. This led to the fact that a large number of designers started to design optical components and systems although they are not specialist in optical design. They use more and more sophisticated optical design and optimization software. The evolutionary algorithms nicely fit here because they do not require greater knowledge of the optical system being optimized. Analyzing the trends of modern optical design it is possible to predict that the optical design programs will be integrated with the CAD/CAM systems and the manufacturing machines. The cycle time between concept, computer model, optimization, machine design and prototype or even custom finished product will likely shrink from months to hours.

The evolutionary algorithms are certain to play an increasingly important role in a range of engineering problems, including design. It seems that optical design in particular offers an ideal context for demonstration of these algorithms: the system modelling is elementary, yet the terrain in the merit function space seems tortuous. The most significant contributions of evolutionary algorithms in the field of optical design are probably to be made in areas where there is relatively little accumulated experience to guide optical designers. Relatively uncharted configuration spaces include systems incorporating inhomogeneous media, diffractive optics or optical systems without symmetries.

The future trends in the development of classical and modern optimization methods in the optical design are the following:
- incorporating good features of the classical and the evolutionary algorithms in one universal algorithm for optical design;
- developing parallel algorithms for optical design on parallel processing computers.

It is a well established fact that the classical algorithms in the optical system optimization are based on numerical methods that are very fast in finding the local minimum but also dependent on the selection of a starting point. Numerical methods always lead to the first local minimum and cannot escape from it. On the other hand, the evolutionary algorithms are global optimization algorithms which mean that they can escape from the local minimums and search the optimization space more thoroughly. They are also not dependent on the starting positions. All these good sides of evolutionary algorithms have one bad side: an average optimization with evolutionary algorithms lasts at least an order of magnitude longer than the optimization with the classical algorithms. It is obvious to try to put good features from both algorithms in one hybrid algorithm. Evolutionary algorithms will search the optimization space for good starting points for the classical algorithm which will then quickly lead to the local minimum. One of local minimums will be also the global minimum. The author expects that the time of execution will be shorter because the final search is done by the classical algorithm.

Modern computer architectures turn to the parallel processing more and more and the evolutionary algorithms can show their full strength only running on a parallel processing computer. Simultaneous running of several evolutionary algorithms allowing migration between the populations of these algorithms can be done on parallel processors. Also the merit functions of individuals in a population can be simultaneously calculated, which tremendously speeds up the convergence of evolutionary algorithms. There is probably no problem to execute the above described hybrid algorithm on a parallel processing computer.

There are many open questions, and there is much important work to be done in the field of optical system optimization. Readers onward!

Appendix A

Description of ADOS – the program for automatic design of optical systems

The ADOS (Automatic Design of Optical Systems) program is a proprietary optical design program. Its development started in 1987 with the following reasons:

- research in classical and modern optimization methods and their application in optimization of optical systems;
- development of a modern optical design program based on PC computers. At that time the state of the art optical design programs were defined by large expensive and user-unfriendly programs running on the also large, expensive mainframe computers. Most of those programs were proprietary and they were not commercially available;
- optical design program that can be easily modified to accept new methods of calculation and optimization for classical optical systems and calculation of new types of optical systems. This possibility is very important when the program is used for design of complex optical and optoelectronical systems for military use.

Before starting the development of the program the author considered various hardware and software solutions that were available to him. The most important factors in analyzing hardware resources are:

- price/performance ratio which is very good for PC computers because of their low price in accordance with other types of computers (engineering graphical workstations, mainframes and supercomputers).
- speed, which is for optical design programs, characterized by "ray surfaces per second" (RSS). The RSS is the number of rays a computer can trace from one surface of an optical system to the next in one second. The first PC computers in 1984 used a processor capable of about 1000 RSS. This was slower than mainframes of the day but PC computers have been developing very fast and modern PC computers using processors from the Intel Pentium family do more than 1 000 000 RSS.
- availability of computer and software development tools for the chosen computer.

After considering all hardware alternatives, the author decided to develop the program on PC computer. At that time the operating system was the MS DOS and later when the MS Windows was developed the program is redesigned to use the graphical user interface from the Windows operating system. From the start the author's intention was to write a user-friendly program for a complete design and optimization of optical systems. The idea was to make integrated environment from which the user could easily perform any action that is needed for design or optimization of optical systems. The author's intentions were the following:

- all commands are available in menus so that the user can easily choose it;
- the program is error proof, which means that no matter what the user does, the program would display error message and help to perform this command correctly.

This also means that user need not memorize a complex sequence of commands any more. The ADOS program has the following features:

- the design and the optimization of various types of optical systems such as different types of objectives, eyepieces, afocal systems, projectors, condensors, collimators;
- the different ways of displaying data in tabular or graphical form;
- the support for the glass database from well known producers of glasses such as Schott;
- the connectivity between the ADOS, where the user can optically design a system to have minimum aberrations, and the AutoCAD program for further optomechanical design. This is acomplished by forming the script files that can be executed in the AutoCAD.

Once an optical system has been entered into the computer, the user can perform various analyses to determine system quality. Usually, the first thing to do is to trace paraxial rays to determine the first-order properties of the optical system. These are the properties describing: focal length, image distance, numerical aperture, magnification, pupil sizes and locations. It is sometimes useful at this point to perform an analysis of the third-order aberrations of the optical system (also known as Seidel aberrations or Seidel coefficients), which gives the designer a preliminary idea of whether or not the chosen optical system will have a chance of meeting its specifications. For all other analyses it is necessary to calculate the ray scheme i.e. a set of well defined rays that are passing through the optical system.

The ADOS program incorporates many of the standard methods of analysis that have been used for decades. The traditional ways of describing the image quality of an optical system include:

- the calculation of five principal monochromatic aberrations: sferical aberration, coma, astigmatism, field curvature and distortion;
- the calculation of transverse ray abberations;
- the calculation of chromatic abberations;
- the calculation of optical path difference;
- the calculation of wave aberrations;
- the calculation of spot diagrams and through-focus spot diagrams for finding the best image position;

With the aid of numerical integration and the fast Fourier transforms (FFTs), a host of additional tools for the analysis are available such as the modulation transfer function (MTF) which describe the relative image contrast as a function of increasing spatial resolution. The ADOS program provides many types of the MTF analysis, including the MTF as a function of field position, through the focus MTF for finding the best image position and the diffraction MTF.

The most important addition to lens-design programs have been the optimization routines, in which the starting optical system is modified in order to reduce aberrations. The goal of optimization is to take a starting optical system and change it to improve its performance (the starting optical system should have a necessary number of optical surfaces of sutable types, since optimization can change only the values of parameters not the number or type of surfaces). Optical systems are defined by the parameters associated with the individual surfaces of the optical system. These parameters can be divided in two main groups:

— the basic parameters which must be defined for each optical system;
— the optional parameters which are defined only if these parameters exist, for example if the surface is aspheric, the user must define the aspherical constants.

The basic parameters are the following parameters:

— the radius of curvature for each surface of an optical system;
— the separation between two surfaces of an optical system;
— the glasses of which components of an optical system are made;
— the clear radius for each surface of an optical system.

To completely define and analyze the optical system, the user must, besides basic parameters, define the ray scheme, which describes rays which will be all traced through the optical system. The user can also define properties of the optical system (for example, radii of curvature, element thickness, glass types, and air spaces) that are allowed to vary in optimization. The optimization algorithm can optimize the optical system by changing the variables and minimizing the merit function.

Because the optimization of the merit function will find a local minimum, it is up to the designer to bring the starting optical system to a suitable quality before the optimization so that the local minimum is close to the global minimum. In other words, it is not possible to choose five plane – parallel plates of arbitrarily chosen glass and expect the program to find an optimum solution.

There have been recent advances in the development of global optimization techniques but the user must be careful with them. Although the global optimization can be a useful tool in the lens-designer's toolbox, it will not satisfy the designer's wish to find the best optical system automatically. In general, the global optimization can produce numerous starting points from which the designer can locally optimize.

A.1   Limitations of the program and the comparison with other commercially available programs for optical design

The ADOS program is developed as a result of research in classical and new-global optimization methods in optical design and as a tool for solving design problems of the complex optical and optoelectronical systems. It is a product of research and development of only one man for more then ten years.

Now at the open market there are four types of optical engineering softwares that cover various different parts of optical engineering:

— The optical design programs also known as classical ray tracers. They alow an optical system to be designed, analyzed, optimized and toleranced. They are ideal for designing various objectives, telescopes and microscopes because they provide powerful optimization methods with strong image analysis features. Typical programs in this category are the CODE V from the Optical Research Associates, the OSLO developed by the Sinclair Optics and now distributed by the Lambda Research Corporation, the ZEMAX from the Focus software.

— The non-sequential ray tracers which are used for modeling optical systems in which light can follow multiple paths, or in which light suffers multiple reflections. A typical representative of this kind of programs is the OptiCAD from the OptiCAD Corporation.

— The optical physics programs that represent light as an elecromagnetic field with the amplitude and phase. They are necessary when the underlying physics of a problem must be understood. Typical applications for this kind of program are laser design, photolitography, diffractive optics for beam control. A representative program of this group is the GLAD from the Applied Optics Reseach.

— The thin film design programs are used for designing optical coatings which are essential for a good optical system. The typical representatives are the Essential Macleod from the Thin Film Center and the FilmStar from the FTG Software Associates.

The ADOS program can be compared only with programs from the first group. It is very important to notice that almost all programs in this group are results of very long development process (usually longer than twenty years) and large teams of researchers and developers-programmers (usually more than ten). They cover almost everything that one optical designer needs in his regular work.

The ADOS program in comparison to those programs has many restrictions and limitations resulting from the following factors:

— The ADOS program has only features that are neccessary for day-to-day work of a standard (conventional) optical designer. The features neccessary for development of other kinds of optical systems are not included in the program. For example, the program supports only the spherical and the aspherical optical surfaces and does not support the diffraction gratings or the gradient index optics.

– Short time of development in comparison to other commercial programs and only one researcher severely restricted the program development so that whole fields in optical design are not included such as the tolerancing of optical systems, various physical optics calculations such as point spread function, line spread function, wavefront analysis, polarization and nonseuential ray trace.

In the field of optimization of optical systems the ADOS program can be compared to any other optical design programs. It has unique global optimization methods such as the adaptive steady state genetic algorithm and evolution strategies that are not available in other optical design programs.

The tabular review of features of the ADOS and two commercially available programs (ZEMAX and OSLO) are given in Table A.1.

Table A.1. The comparative analysis of the following lens design programs: ADOS, ZEMAX and OSLO

		ADOS	ZEMAX	OSLO
Surface types	Spherical	Yes	Yes	Yes
	Aspherical	Yes	Yes	Yes
	Cylindrical	No	Yes	Yes
	Toroidal	No	Yes	Yes
	Diffraction gratings	No	Yes	Yes
	User defined	No	Yes	Yes
Ray tracing	Paraxial	Yes	Yes	Yes
	Seidel	Yes	Yes	Yes
	Bucdahl	No	Yes	Yes
	MIL ray trace	Yes	Yes	Yes
	aberration calculation	Yes	Yes	Yes
Analysis	aberrations	Yes	Yes	Yes
(graphics)	OPD	Yes	Yes	Yes
	Spot diagram	Yes	Yes	Yes
	Best image position	Yes	Yes	Yes
	Geometric MTF	Yes	Yes	Yes
	Diffraction MTF	Yes	Yes	Yes
	Gaussian beam	No	Yes	Yes
	3D Wavefront	No	Yes	Yes
	Point spread function	No	Yes	Yes
	Line spread function	No	Yes	Yes
Optimization	Dumped DLS	Yes	Yes	Yes
Global	Simple GA	No	Yes	No
	Adaptive Simulated Annealing	No	No	Yes
	ASSGA	Yes	No	No

		ADOS	ZEMAX	OSLO
	EVOL ES	Yes	No	No
	GRUP ES	Yes	No	No
	REKO ES	Yes	No	No
	KORR ES	Yes	No	No
Tools	Export to CAD	Yes	Yes	Yes
	Tolerancing	No	Yes	Yes
	Polarization	No	Yes	Yes
	Illumination Analysis	No	Yes	Yes
	Zoom lenses	No	Yes	Yes

All further information about the ADOS program can be obtained from the author on the email address darko@afrodita.rcub.bg.ac.yu and on the web site http://www.diginaut.com/shareware/ados/. The ADOS program can be downloaded from the web site http://www.diginaut.com/shareware/ados/download.htm.

Online bibliography

There is a large number of online resources concerning evolutionary computation. On of the most interesting and also the most complete bibliographies is collected by Jarmo Alander from the University of Wasa in Finland. He collected a large number of references concerning all aspects of the evolutionary computation and their implementation in various fields of science and technology. At the end of year 2000 the database with references was quite large, more than 13600 references. All references are collected in several reports concerning only one aspect of the evolutionary computation. Among the others there are the following reports:

- basics of genetic algorithms;
- implementation of genetic algorithms;
- theory and analysis of genetic algorithms;
- evolution strategies;
- genetic algorithms in computer sciences;
- genetic algorithms in computer aided design;
- genetic algorithms in artificial intelligence;
- genetic algorithms in chemistry and physics;
- genetic algorithms in engineering;
- genetic algorithms and optimization.

All reports are available online from the anonymous ftp site: `ftp.uwasa.fi` in the directory `cs/report94-1`.

Probably the most interesting is the report, which shows the implementation of genetic algorithms in optics and image processing. The title of this report is: "An Indexed Bibliography of Genetic Algorithms in Optics and Image Processing" and it is like other reports available online from the anonymous ftp site `ftp.uwasa.fi` in the directory `cs/report91-1` file `gaOPTICSbib.ps.Z`. The report is a standard postscript file compressed with the Unix compress tool by using the adaptive Lempel – Ziv coding.

References

[1] S. Rosen, C. Eldert, Least – squares method for optical correction, J. Opt. Soc. Am. vol. 44 no. 3, pp.250 – 252, 1954.

[2] T. H. Jamieson, *Optimization techniques in lens design*, Monographs on applied optics no. 5 New York: American Elsevier Publishing Company Inc., 1971.

[3] M. Kidger: The application of electronic computer to the design of optical systems including aspheric lenses, Ph. D. Thesis, University of London, 1971.

[4] C. G. Wynne, P. M. J. H. Wormell, Lens design by computer, App. Opt. vol. 2, no. 12, pp. 1233 – 1238, 1963.

[5] R. R. Mayer, "Theoretical and computational aspects of non linear regression", in *Nonlinear programming*, J.B. Rosen, O. L. Mangasarian, K. Ritter eds. Academic Press 1970.

[6] J. Meiron, Damped least squares method for automatic lens design, J. Opt. Soc. Am. vol. 55, no. 9, pp. 1105 – 1109, 1965.

[7] D. P. Buchele, Damping factor for the least squares method of optical design, App. Opt. vol. 7, no. 12, pp. 2433 – 2435, 1968.

[8] D. C. Dilworth, Pseudo – second derivative matrix and its application to automatic lens design, App. Opt. vol. 17, no. 21, pp. 3372 – 3375, 1978.

[9] D. C. Dilworth, Improved convergence with pseudo – second derivative (PSD) optimization method, SPIE, vol. 399, pp. 159 – 165, 1983.

[10] D. C. Dilworth: Automatic lens optimization: recent improvements, SPIE vol. 554, pp. 191 – 196, 1985.

[11] E. D. Huber, Extrapolated least squares optimization a new approach to least squares optimization in optical design, App. Opt. vol. 21, no. 10, 1705 – 1707, 1982.

[12] H. Matsui, K. Tanaka, Determination method of an initial damping factor in the damped least squares problem, App. Opt. vol. 33, no. 13, pp. 2411 – 2418, 1994.

[13] P. N. Robb, Accelerating convergence in automatic lens design, App. Opt. vol. 18, no. 24, pp. 4191 – 4194, 1979.

[14] G. H. Spencer, Flexible lens correction procedure, App. Opt. vol. 2, no. 12, pp. 1257 – 1264, 1963.

[15] D. S. Grey, Aberration theories for semiautomatic lens design by electronic computers I Preliminary remarks, J. Opt. Soc. Am. vol. 53, no. 6, pp. 672 – 676, 1963.

[16] D. S. Grey, Aberration theories for semiautomatic lens design by electronic computers II a specific computer program, J. Opt. Soc. Am. vol. 53, no. 6, pp. 677 – 680, 1963.

[17] D. S. Grey, R. J. Pegis, T. P. Vogl, A. K. Rigler, The generalized orthonormal optimization program and its applications, in *Recent advances in optimization techniques*, A. Lavi, T. P. Vogl eds., New York: John Wiley & Sons, Inc. 1966

[18] N. Metropolis, A. Rosenbluth, A. Teller, E. Teller, Equations of state calculations by fast computing machines, J. Chem. Phys. vol. 21, pp. 1087 – 1091, 1953.

[19] S. Kirkpatrick, C. Gelatt, M. Vecchi, Optimization by stimulated annealing, Science, vol. 220, no. 4598, pp. 671 – 679, 1983.

[20] C. Reeves, Ed.: Modern heuristic techniques for combinatorial problems, London: McGraw – Hill Book Company, 1995.

[21] H. P. Schwefell, *Evolution and optimum seeking*, New York: John Wiley & Sons Inc., 1995.

[22] G. Hearn, Practical use of generalized simulated annealing optimization on microcomputers, SPIE vol. 1354, 1990.

[23] E. Glatzel, R. Wilson, Adaptive automatic correction in optical design, Appl. Opt. vol. 7 no. 2, pp. 265 – 276, 1968.

[24] J. Rayces, Ten years of lens design with Glatzel's adaptive method, SPIE vol. 237, pp. 75 – 84, 1980.

[25] C. Darwin, *On the origin of species by means of natural selection or the preservations of favoured races in the struggle for life*, London: John Murray, 1859.

[26] J. Holland, *Adaptation in natural and artificial systems*, 2nd edition, Cambridge, Massachusetts: MIT Press, 1992.

[27] K. De Jong, An analysis of the behaviour of a class of genetic adaptive systems, Ph. D. Thesis, University of Michigan, 1975.

[28] D. E. Goldberg, *Genetic algorithms in search, optimization & machine learning*, Reading, Massachusetts: Addison – Wesley Publishing Co. Inc., 1989.

[29] Z. Michalewicz, *Genetic algorithms + data structures = evolution programs*, 2nd extended edition, Berlin: Springer – Verlag, 1994.

[30] J.J. Grefenstette, J. E. Baker, "How genetic algorithms work: a critical look at implicit parallelism", in *Proceedings of the 3rd international conference on genetic algorithms and their applications*, J. D. Schaffer, ed. San Mateo, California: Morgan Kaufman Publishers, 1989.

[31] T. Bäck, *Evolutionary algorithms in theory and practice*, New York, Oxford: Oxford University Press, 1996.

[32] F. Z. Brill, D. E. Brown, W. N. Martin, Fast genetic selection of features for neural network classifiers, IEEE Trans. on neural networks, vol. 3, no. 2, pp. 324 – 328, 1992.

[33] R. Tanese, Distributed genetic algorithms for function optimization, Ph. D. Thesis, University of Michigan, 1989.

[34] L. Davis, *Handbook of genetic algorithms*, New York: Van Nostrand Reinhold, 1991.

[35] J. Baker, "Reducing bias and inefficiency in the selection algorithm", in *Proceedings of the 2nd international conference on genetic algorithms and their applications*, J. Grefenstette, ed. Hillsdale NJ: Lawrence Erlbaum Associates, 1987.

[36] D. Goldberg, A note on Boltzman tournament selection for genetic algorithms and population – oriented simulated annealing, Complex Systems, vol. 4, pp. 445 – 460, 1990.

[37] M. De la Maza, B. Tidor, "An analysis of selection procedures with particular attention paid to proportional and Boltzmann selection", in *Proceedings of the 5th international conference on genetic algorithms,* S. Forrest, ed. San Mateo, California: Morgan Kaufman Publishers, 1993.

[38] J. Baker, "Adaptive selection methods for genetic methods for genetic algorithms" in *Proceedings of the 1st international conference on genetic algorithms,* J. Grefenstette, ed. Hillsdale NJ: Lawrence Erlbaum Associates, 1985.

[39] D. Goldberg, K. Deb, "A comparative analysis of selection schemes used in genetic algorithms" in *Foundations of genetic algorithms,* G. Rawlins, ed. San Mateo, California: Morgan Kaufman Publishers, 1991.

[40] D.Whitley, "Genitor: a different genetic algorithm" in *Proceedings of the 3rd international conference on genetic algorithms,* J. D. Shaffer, ed. San Mateo, California: Morgan Kaufman Publishers, 1989.

[41] G. Syswerda, "Uniform crossover in genetic algorithms", in *Proceedings of the 3rd international conference on genetic algorithms,* J. D. Shaffer, ed. San Mateo, California: Morgan Kaufman Publishers, 1989.

[42] J. Grefenstette, Optimization of control parameters for genetic algorithms, IEEE Trans. on Systems, Man and Cybernetics, vol. 16, no. 1, pp. 122 – 128, 1986.

[43] J. Schaffer, R.Caruana, L. Eshelman, R. Das, "A study of control parameters affecting online performance of genetic algorithms for function optimization" in *Proceedings of the 3rd international conference on genetic algorithms,* J. D. Schaffer, ed. San Mateo, California: Morgan Kaufman Publishers, 1989.

[44] B.P. Demidovich, I. A. Maron, *Computational mathematics,* Moscow: Mir Publishers, 1981.

[45] N. V. Kopchenova, I. A. Maron, *Computational mathematics worked examples and problems with elements of theory,* Moscow, Mir Publishers: 1984.

Index

The manufacturer's authorised representative in the EU is Springer
Nature Customer Service Centre GmbH, Europaplatz 3, 69115 Heidelberg,
Germany. If you have any concerns regarding our products, please
contact ProductSafety@springernature.com

Printed and bound by CPI Group (UK) Ltd, Croydon, CR0 4YY
23/04/2026
02095624-0003